Contributions to Management Science

The series *Contributions to Management Science* contains research publications in all fields of business and management science. These publications are primarily monographs and multiple author works containing new research results, and also feature selected conference-based publications are also considered. The focus of the series lies in presenting the development of latest theoretical and empirical research across different viewpoints.

This book series is indexed in Scopus.

María Teresa Del Val Núñez •
Alba Yela Aránega • Domingo Ribeiro-Soriano
Editors

Artificial Intelligence and Business Transformation

Impact in HR Management, Innovation and Technology Challenges

Springer

Editors
María Teresa Del Val Núñez
University of Alcalá
Alcalá de Henares, Madrid, Spain

Alba Yela Aránega
University of Alcalá
Alcalá de Henares, Madrid, Spain

Domingo Ribeiro-Soriano
University of Valencia
Valencia, Spain

ISSN 1431-1941 ISSN 2197-716X (electronic)
Contributions to Management Science
ISBN 978-3-031-58703-0 ISBN 978-3-031-58704-7 (eBook)
https://doi.org/10.1007/978-3-031-58704-7

This Springer imprint is published by the registered company Springer Nature Switzerland AG
The registered company address is: Gewerbestrasse 11, 6330 Cham, Switzerland

Foreword

In contemporary society, the pervasive influence of rapid technological advancements necessitates prompt adaptation to evolving environments and imposes novel competency requirements upon individuals. This era of digital proliferation, underpinned by the collaborative efforts of experts, scholars, policymakers, scientists, and organizational leaders, has ushered in unprecedented opportunities for the advancement of corporate departments.

At the forefront of this paradigm shift stands artificial intelligence (AI), a ubiquitous force permeating various domains. Over recent years, the integration of AI has emerged as a pivotal catalyst for organizational growth, spearheading the digital transformation initiatives of enterprises worldwide. While extant literature predominantly examines AI through the lens of machine learning applications, commensurate attention must be devoted to elucidating the challenges arising from its practical implementation and the nascent nature of strategic AI utilization for driving business value creation.

In response to the imperatives of change, companies have embraced innovative strategies, infusing creativity into their strategic decision-making processes to navigate dynamic landscapes successfully. Consequently, business models have undergone profound metamorphoses, striving to enhance competitiveness through strategic recalibrations encompassing organizational design, scale, and core operations. This transformative trajectory has engendered novel considerations regarding the future landscape of work, encompassing the creation of new roles, the imperative to cultivate digital competencies, and the reimagining of team dynamics. Central to this discourse is a nuanced exploration of the human resources (HR) domain, underscored by the profound impact of digitalization on the workforce. Indeed, employees occupy a pivotal role in organizational sustenance and evolution, necessitating an in-depth examination of HR dynamics amidst the backdrop of technological upheaval.

This volume endeavors to scrutinize artificial intelligence from diverse vantage points, charting its manifold impacts and trajectories within the organizational milieu. Structurally, the book is bifurcated into two cardinal sections, each addressing distinct facets of AI's influence on organizational dynamics. The initial segment elucidates strategies for business transformation, delineating novel pathways for augmenting productivity and profitability. In contrast, the subsequent section delves into the integration of AI within the HR domain, probing both its tangible and intangible ramifications. This holistic inquiry encompasses considerations spanning the creation of new job roles, the imperative to cultivate digital proficiencies, and the imperative to foster innovative paradigms of team management.

As the Vice-President for International Relations at the University of Alcalá, I fervently advocate for scholarly endeavors such as this publication, which epitomizes the commitment to excellence in academic discourse. The richness of content encapsulated within this volume, alongside its diverse array of contributions imbued with global perspectives, augments its scholarly merit and underscores its significance in advancing the frontiers of knowledge.

University of Alcalá
Madrid, Spain

Julio Cañero Serrano

Artificial Intelligence and Business Transformation

Artificial Intelligence and Business Transformation is a book series composed by the works of professionals where the new trends in this field are contextualized, deepened, and defined. The rapid evolution of the development of artificial intelligence, focused on the field of business development, and more specifically on human resources, reflects new business strategies, innovative tools in business and team management, and current practices in human resources departments. This book is mainly written by professionals in the field, consultants and academics who seek to capture the reality, as well as to generate discussions and reflection on the topic.

Contents

Part II Human Resources

About the Editors and Contributors

About the Editors

María Teresa del Val Núñez received a PhD in Economics and Business Administration from the University of Alcalá, Spain, in 1993. She is currently Full Professor of Business and Management at the University of Alcalá, General Director of FGUA, Alcalingua, and CRUSA. She has previously received a DAAD scholarship (1988–1990) and has worked as a collaborating researcher at the IfM Institut für Mittelstandsforschung in Bonn, Germany. She has actively participated in various EU, national, and international research projects. Her research has led her to participate in numerous national and international seminars. She is also the author of several books and articles published in national and international journals. She is an honorary member of the Alexander von Humboldt Association in Spain, and in 2014, she had the distinction of receiving a Silver Medal from the University of Alcalá.

Alba Yela Aránega is a PhD in Economics and Business Management at the University of Alcalá, Spain. She is currently Business Organisation and Management Assistant Professor at the University of Alcalá. She is a reviewer of some international reviews and handling editor in international journals in SSCI-ranked journals. She has published widely on topics such as mindfulness for intraentrepreneurship, labor integration of young people, human resources, and training.

Domingo Ribeiro-Soriano is Professor of Business Administration at the Universitat de València, Spain. He is also the co-director of the "Entrepreneurship: from student to Entrepreneur" Chair. As a researcher, he has published more than 130 papers in SSCI-ranked journals. Throughout his career, he has edited and contributed to books, journals, and conferences and has delivered keynote speeches at international conferences. He has also led several EU-funded projects and contracts with private companies. Before starting his career in academia, he worked as a consultant at EY (formerly Ernst & Young).

Contributors

Craig Graham Anderson Stirling Management School, University of Stirling, Stirling, UK

Katerina Beta School of Strategy and Leadership, Coventry University, Coventry, UK

Simone Bevilacqua Department of Management, Università degli Studi di Torino, Turin, Italy

Gülin İdil S. Bolatan Industrial Engineering Department, Alanya Alaaddin Keykubat University, Alanya, Turkey

Stefano Bresciani Department of Management, Università degli Studi di Torino, Turin, Italy

M. Ángeles López Cabarcos Department of Business Administration, University of Santiago de Compostela, Santiago, Spain

Rafael Castaño Sánchez Department of Economics and Business, Faculty of Economics, Business and Tourism, University of Alcalá, Madrid, Spain

Raúl Castaño Urueña Department of Economics and Business, Faculty of Economics, Business and Tourism, University of Alcalá, Madrid, Spain

Andy Coleman School of Strategy and Leadership, Coventry University, Coventry, UK

Ricardo Costa-Climent Uppsala University, Uppsala, Sweden

University of Economics and Human Sciences, Warsaw, Poland

Fernando Javier Crecente-Romero Department of Economics and Business Administration, Institute of Economic and Social Analysis (IAES), University of Alcalá, Madrid, Spain

Pedro Cuesta-Valiño University of Alcalá, Faculty of Economics, Business and Tourism, Department of Economics and Business, Madrid, Spain

Tugrul Daim Mark O. Hatfield Cybersecurity & Cyber Defense Policy Center, Portland State University, Portland, OR, USA

Antonio de Lucas Ancillo University of Alcalá, Faculty of Economics, Business and Tourism, Department of Economics and Business, Madrid, Spain

Francisco del Olmo-García Department of Economics and Business Administration, Institute of Economic and Social Analysis (IAES), University of Alcalá, Madrid, Spain

María Teresa Del Val Núñez Department of Economics and Business Administration, Institute of Economic and Social Analysis (IAES), University of Alcalá, Madrid, Spain

Debora Di Caprio Department of Economics and Management, University of Trento, Trento, Italy

Christian Di Prima Department of Management, Università degli Studi di Torino, Turin, Italy

Faculty of Social Sciences and Solvay Business School, Vrije Universiteit Brussels, Brussels, Belgium

Alberto Ferraris Department of Management, Università degli Studi di Torino, Turin, Italy

Gnosis: Mediterranean Institute for Management Science, School of Business, University of Nicosia, Nicosia, Cyprus

José Andrés Gómez Gandía Faculty of Economics, Business and Tourism, Department of Economics and Business, University of Alcalá, Alcalá de Henares, Spain

Rubén Garrido-Yserte Instituto Universitario de Análisis Económico y Social (IAES), Facultad de Ciencias Económicas, Empresariales y Turismo, University of Alcalá, Alcalá de Henares, Spain

Sorin Gavrila Gavrila University of Alcalá, Faculty of Economics, Business and Tourism, Department of Economics and Business, Madrid, Spain

Isaac J. González-López University of Santiago de Compostela, Santiago, Spain

Cristina Blanco González-Tejero Faculty of Economics, Business and Tourism, Department of Economics and Business, University of Alcalá, Alcalá de Henares, Spain

Samet Batuhan Güven Management Engineering Department, Alanya Alaaddin Keykubat University, Alanya, Turkey

Darek M. Haftor Uppsala University, Uppsala, Sweden

University of Economics and Human Sciences, Warsaw, Poland

Blanca García Henche University of Alcalá, Faculty of Economics, Business and Tourism, Department of Economics and Business, Madrid, Spain

Andreas M. Hilger Chair of Leadership and Organization, University of Regensburg, Regensburg, Germany

Sergey Kazakov University of Alcalá, Faculty of Economics, Business and Tourism, Department of Economics and Business, Madrid, Spain

Raquel Marín Instituto Complutense de Estudios Internacionales, Universidad Complutense de Madrid, Madrid, Spain

Departamento de Análisis Económico y Economía Cuantitativa, Universidad Complutense de Madrid, Madrid, Spain

Ángel Javier Álvarez Miguel Polytechnics School, Department of Computer Science, University of Alcalá, Alcalá de Henares, Spain

Sangeeta Nar Chair of Innovation and Technology Management, University of Regensburg, Regensburg, Germany

Estela Núñez-Barriopedro University of Alcalá, Faculty of Economics, Business and Tourism, Department of Economics and Business, Madrid, Spain

Óscar Montes Pineda Instituto Universitario de Análisis Económico y Social (IAES), Facultad de Ciencias Económicas, Empresariales y Turismo, University of Alcalá, Alcalá de Henares, Spain

Juan Piñeiro-Chousa Department of Finance and Accounting, University of Santiago de Compostela, Santiago, Spain

Kerstin Rego Chair of Leadership and Organization, University of Regensburg, Regensburg, Germany

Arturo Orea Rocha University of Alcalá, Faculty of Economics, Business and Tourism, Department of Economics and Business, Madrid, Spain

Noelia Romero-Castro Department of Finance and Accounting, University of Santiago de Compostela, Santiago, Spain

Francisco J. Santos-Arteaga Departamento de Análisis Económico y Economía Cuantitativa, Universidad Complutense de Madrid, Madrid, Spain

María Sarabia-Alegría Department of Economics and Business Administration, Institute of Economic and Social Analysis (IAES), University of Alcalá,, Madrid, Spain

Raquel Sastre Instituto Universitario ESEADE, Buenos Aires University (UBA), Faculty of Economical Sciences, Buenos Aires, Argentina

Christian Scharff Chair of Leadership and Organization, University of Regensburg, Regensburg, Germany

Marcin W. Staniewski University of Economics and Human Sciences, Warsaw, Poland

Madjid Tavana Business Systems and Analytics Department, La Salle University, Philadelphia, PA, USA

Business Information Systems Department, Faculty of Business Administration and Economics, University of Paderborn, Paderborn, Germany

José Ángel Tébar-Sáez University of Alcalá, Faculty of Economics, Business and Tourism, Department of Economics and Business, Madrid, Spain

Alba Yela Aránega Department of Economics and Business, Faculty of Economics, Business and Tourism, University of Alcalá, Madrid, Spain

List of Figures

The Behavioural Science of Using AI in HRM Decision-Making: When It Helps and When It Goes Wrong

List of Tables

Part I
Business Transformation

New Lines of Business Development: Artificial Intelligence in Business

José Andrés Gómez Gandía, Cristina Blanco González-Tejero, and Ángel Javier Álvarez Miguel

Abstract This chapter provides an analysis of the different doctrines and learnings framed at the intersection of Artificial Intelligence (AI) and business development. Thus, it explores how AI has evolved to drive the creation of new business opportunities in the dynamic international environment. In the first section, we define AI and its evolutionary trajectory in the context of new business. The fundamentals of AI in business development are broken down in the second section, covering the essential concepts of AI as applied to business management and how this technology empowers the identification of emerging opportunities and trends in the marketplace. Section three examines specific applications of AI in the creation of new lines of business, focusing on the identification of market opportunities. As we move forward, the fourth section explores the strategic integration of AI into the business environment. The importance of a business culture open to innovation and technology is highlighted, along with the implementation of digital transformation strategies to facilitate the effective adoption of AI. The analysis progresses to the fifth section, where the future of AI in business development is projected, highlighting technological advances and the continued potential of AI to create new business opportunities. Finally, the sixth section draws conclusions, recapping the fundamental role of AI in the development of new lines of business, as well as the implications and challenges facing companies and entrepreneurs in this era of AI-driven transformation. Taken together, this analysis highlights how AI has emerged as an essential tool for innovation and business strategy, triggering an ever-evolving landscape.

J. A. G. Gandía (✉) · C. B. González-Tejero
Faculty of Economics, Business and Tourism, Department of Economics and Business, University of Alcalá, Alcalá de Henares, Spain
e-mail: josea.gomez@uah.es; cristina.blancog@uah.es

Á. J. Á. Miguel
Polytechnics School, Department of Computer Science, University of Alcalá, Alcalá de Henares, Spain
e-mail: a.alvarez@uah.es

M. T. Del Val Núñez et al. (eds.), *Artificial Intelligence and Business Transformation*, Contributions to Management Science,
https://doi.org/10.1007/978-3-031-58704-7_1

1 Introduction

The pursuit of new lines of business development has been significantly shaped by the transformative potential of artificial intelligence (AI) (Codini et al., 2023; Miller, 2018). This introductory exposition delves into the evolutionary trajectory of AI within the realm of emerging business paradigms, elucidating its foundational concepts and its burgeoning influence in identifying emerging market opportunities and trends. As AI increasingly permeates diverse industries (Dwivedi et al., 2021), its integration is of paramount importance not only to expand organizational adaptability, but also to generate strategic synergy between technological advances and business innovation. This exploration attempts to navigate the dynamic landscape in which the evolution of AI converges with the propulsive contours of new business possibilities, highlighting the intricate interplay between technological augmentation and potential business horizons.

AI is composed of a very broad technology, making it very difficult to provide a concrete definition. Therefore, many considerations arise, one of the most common being the idea that AI is based on the simulation, by computers, of human behavior. This makes it possible to train machines to acquire skills related to decision making, judgment and emulation of human behavior (Da Xu et al., 2021). Within the field of AI, we find a wide range, from specialized AI, which focuses on the execution of specific and limited tasks, to general AI, which aims to match human ability to solve a variety of problems (D. Zhang, 2023).

AI is used in various fields, such as economic, social and healthcare (Iqbal et al., 2016) and its operation is based on the application of different techniques. As a specific approach within the field of AI, it is worth mentioning machine learning provides the machine with the ability to learn (Bishop & Nasrabadi, 2006).Cognitive ability is introduced in organizations to take decisions that increase productivity and help HR with employee retention (Shrestha et al., 2021) based on data analysis looking for patterns that allow the modification of employee behavior. We also find deep learning (Heaton, 2018) which is framed within machine learning and which uses machine learning algorithms inspired by the neurons of the human brain surpassing human capabilities (Silver et al., 2018). However, these benefits also come at a price, as there are several challenges to overcome to successfully implement analytical models in real business environments. These include the appropriate choice between multiple implementation options, bias and drift in data, mitigation of black-box properties, and reuse of preconfigured (as-a-service) models (Janiesch et al., 2021).

Based on the above definitions, we can see how, within companies, those responsible for ICT and data analysis seek to use AI to provide solutions to business problems that companies face daily, but also to seek higher returns from lower costs (Othman et al., 2016) arising in day-to-day business. The fact that AI can automate tasks (Hofmann et al., 2020; Siderska, 2020; Sobczak, 2022) that were previously performed by a series of workers or analyze the data generated by companies in the development of their activities and being able to optimize processes (Aguirre &

Rodriguez, 2017) and generate knowledge makes it an essential tool for the subsistence of the company (Fink et al., 2021).

2 AI Applied to Business Development

Foundations of AI have become key principles for the development and adaptation of enterprises to the changing environment. This exposition embarks on a comprehensive analysis of the rudimentary principles underlying the symbiotic relationship between AI and the development of new enterprises. Delving into the basic concepts of AI as applied to the business domain, this chapter allows one to consider how different cognitive AI paradigms contribute synergistically to the identification, evaluation, and exploitation of market prospects, thereby strengthening the strategic acumen of firms in an era characterized by rapid technological evolution and transformative economic change.

As a discipline of computer science, AI focuses on the creation of systems capable of performing tasks that normally require human intelligence (Miller, 2018). Its application in business has gained great momentum in recent years, transforming the way in which various aspects of management and decision-making are addressed. In the globalized environment, there is a multitude of data available, which, with correct extraction and processing allow organizations to be assisted in their strategic management. This technique is known as data mining, which involves the discovery and extraction of hidden patterns and relationships in massive data sets (Gutiérrez & Molina, 2016). Thus, with the help of AI techniques, companies can identify market trends, customer segments, and growth opportunities that might otherwise go unnoticed. Thus, AI provides evidence-based data analysis, which improves decision-making (Yarlagadda, 2018) and helps optimize complex business processes by analyzing multiple variables and finding optimal solutions (Chi-Hsien & Nagasawa, 2019). Therefore, as technology continues to evolve, the strategic integration of AI can provide significant competitive advantages (Wilson & Daugherty, 2018) in the business world.

The functions of AI have been widely considered by a multitude of researchers (Caner & Bhatti, 2020; Enholm et al., 2022). Within AI, deep learning should be highlighted, which refers to the study of AI algorithms that use multiple processing layers to learn and extract features from data, allowing them to understand more complex concepts and patterns in information analysis and classification applications (Deng & Yu, 2014). The use of AI algorithms as well as machine learning (ML) have been applied in most economic sectors. ML seeks to teach how machines should manage data more efficiently by simulating human learning and certain algorithms can be implemented for such learning. Machine learning algorithms can be used to perform predictive analytics, identify patterns, and make informed decisions based on historical data (Jiang et al., 2017). Natural language processing (NLP) is known as another major area within AI, which focuses on enabling machines to understand, interpret and generate human language naturally (C. Zhang et al., 2020). In the business context, NLP is used to analyze customer feedback, automate online responses, and extract valuable information from large volumes of text.

3 Open Culture for Innovation and Technology

A corporate culture that fosters innovation and technology adoption is not only essential to keep up with rapid changes in the business environment, but also to seize new opportunities and generate significant competitive advantages. In the current era, a company's ability to embrace innovation and technology largely helps define its success.

Today, in a business context characterized by rapid technological evolution and disruptive changes in markets, the importance of fostering a business culture that embraces innovation (Bokrantz et al., 2017) and technology is highlighted. This approach is critical to ensure the success and sustainability of organizations. Those companies with a culture open to innovation and technology can adjust with agility to these changes, maintaining their relevance and competitiveness (see Fig. 1). In addition, an innovation and technology-oriented culture enables companies to identify previously unnoticed business opportunities. Consequently, it allows synergy between the different parts of the organization, giving rise to collaborative work and management spaces. In that sense, the adoption of new technologies, as experts

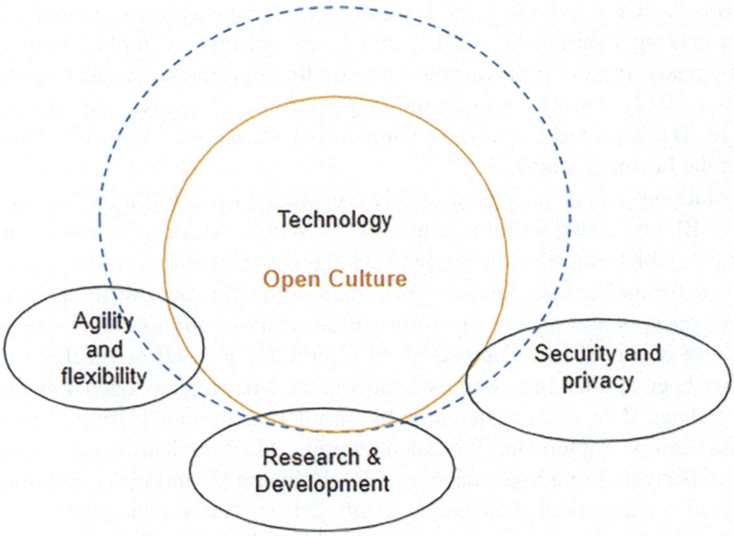

Fig. 1 Digital synergy

have pointed out (Akter et al., 2022), can open the door to new business models and market niches, driving business growth.

In business culture, the interrelationship shown between the different elements is critical to success. An open culture in technology (Leal-Rodríguez et al., 2023) will foster innovation and collaboration, while both agility and flexibility will allow it to adapt to technological changes. The fact that organizations maintain high budgets generates an investment in research and development that will foster a learning environment, developing the organizations' teams, and above all, cultivating a culture that supports this profound process of change by seeking to keep abreast of all technologies (Kane, 2019). We must keep in mind that both security (Cardholm, 2016) and privacy (Scarpi et al., 2022) are key to the ethical use of technology.

Considering the changing environment, to compete in it, one needs to adapt to technological and innovation trends, which is achieved through a culture open to innovation (Monostori et al., 2016). Therefore, as shown in Fig. 1, technology, in addition to contributing to improve operational efficiency, automating processes, optimizing operations, and reducing costs (Zheng et al., 2018), allows agility and flexibility in the organization, generates continuous R&D paths, and provides security in the treatment of information.

4 Implementation of Digital Transformation Strategies for AI Adoption

Digital transformation has become a strategic necessity for companies that want to stay competitive and relevant in today's business landscape. Effective AI adoption can make the difference between a laggard and an innovation leader.

Successful adoption of AI in an organization involves several fundamental steps which can be summarized in a strategic process. First, a thorough assessment of the company's current situation must be conducted (Legner et al., 2017), understanding its technological position, organizational culture, processes, and digital skills. This provides the basis for identifying areas of opportunity where AI can generate the greatest impact (Reim et al., 2020). Once the current situation is understood, clear and measurable objectives should be set that are aligned with the company's strategic vision. These objectives should be realistic in terms of time and available resources.

To accomplish this transformation, the formation of a multidisciplinary team is required, involving professionals from diverse areas (Sharma & Sharma, 2019), including technology, business, marketing, and human resources. Interdepartmental collaboration and a holistic approach are essential for successful AI adoption (Albukhitan, 2020). Talent acquisition and staff training in AI are crucial steps, either through hiring experts or in-house training (Nadeem et al., 2018). In addition, the company's technology infrastructure (Akerkar, 2019) must be assessed and upgraded (Akerkar, 2019) to support AI solutions, which may include the adoption of cloud platforms (Jasiulewicz-Kaczmarek & Gola, 2019), data analytics tools, and specialized hardware.

The implementation of pilot projects in selected areas (Giesen et al., 2010) allows testing and fine-tuning of AI solutions prior to large-scale implementation. These solutions must be carefully integrated into existing business processes (Anagnoste, 2017), often requiring modifications to workflows (Matt et al., 2015) and staff training. Finally, promoting an organizational culture open to innovation and technology (Reim et al., 2020) is essential, communicating the importance of digital transformation and providing support to employees to adapt to these changes. Therefore, this strategic and holistic approach enables careful planning, strategic vision, and collaborative execution for the effective adoption of AI in an organization.

5 Business Ideas

In today's context of highly competitive business and constantly changing market environments, the ability to identify opportunities and emerging trends is crucial to maintain a competitive advantage. Companies are attracted to AI for two main reasons: technological capabilities and cost reduction (T. Davenport et al., 2020). AI has proved to be an invaluable tool in this process by enabling companies to analyze large amounts of data (Cortez & Johnston, 2017) in real time, increasing efficiency (Bag et al., 2021) and discovering patterns that might go unnoticed by traditional methods. Thus, it facilitates the generation of ideas and concepts for new products and services. Through data analysis and scenario simulation, companies can design offerings that address unmet needs in the market.

5.1 The Business Idea Lifecycle and Links to AI

Identifying market opportunities is an essential process in business management, as it allows organizations to discover and exploit niches and unmet needs in the market. Companies can test hypotheses and adjust their approaches according to the results (Mishra & Pani, 2021). Consequently, AI has radically transformed innovation processes (Mughal, 2018), enabling a deeper and more accurate understanding of the competitive environment and consumer behavior.

AI plays a pivotal role in transforming market analysis and business decision-making throughout the product lifecycle (Fig. 2). Thus, Fig. 2 shows different examples in which AI would be useful, taking into account the phase of the product life cycle. Its ability to process and analyze vast volumes of data from diverse sources enables in-depth and rapid analysis that reveals emerging patterns and detailed insights into market preferences and behaviors. In addition, AI enables detailed competitive analysis by monitoring competitors' activities and detecting

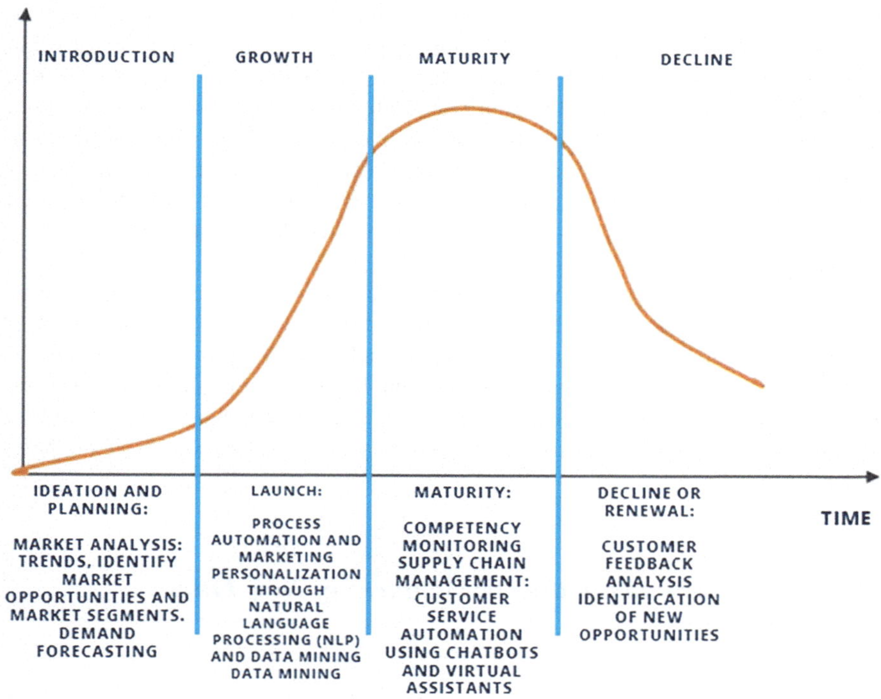

Fig. 2 Life cycle

areas where they could be improved or where they are not meeting market needs fully (Feng et al., 2021).

AI enables the analysis of large volumes of data (Cortez & Johnston, 2017; Kaplan & Haenlein, 2019; Mughal, 2022) from diverse sources, such as social networks, online transactions, and customer feedback, to identify emerging patterns and trends in the marketplace. This helps companies detect (Grewal et al., 2020) unexplored opportunities and anticipate changing consumer needs. They can then tailor their products and services to individual customer preferences, generating greater satisfaction (Chauhan et al., 2022) and loyalty. This real-time analytics capability enables entrepreneurs and businesspeople to identify emerging trends long before they are apparent to conventional methods, so proper management and processing of this data would provide relevant information and competitive advantage. Machine learning and data mining algorithms can identify hidden patterns and relationships in data (Bharadiya, 2023) (Bharadiya, 2023). In the phases of implementation and development of the idea in the market, it is essential to consider personalization and market segmentation. Therefore, by analyzing customer behaviors and preferences, AI can identify opportunities to offer specific products and services that meet individual consumer needs (Rust & Huang, 2014). By processing text, audio, images, and video, AI can analyze online opinions, comments, and reviews to gauge consumer sentiment toward a brand or product (Wang & Kosinski, 2018). This not only helps to identify emerging trends but also to foresee possible changes in public perception. In the later stages of the business lifecycle, it is critical to consider research and development (R&D) for predictive forecasting of future demands to avoid decline. By applying predictive models, AI can anticipate future demands for products and services based on historical patterns and external factors (Ma & Sun, 2020; Valcozzena et al., 2023). This enables companies to be prepared to meet those emerging demands before they occur, thus proactively managing changing needs.

Consequently, AI expands the ability of businesses to identify emerging opportunities and trends by analysing large volumes of data (Dwivedi et al., 2021), detecting complex patterns, personalizing supply, and predicting future demand. By adopting AI-informed strategies, companies can be better prepared to capitalize on these emerging opportunities and stay ahead in a constantly evolving marketplace (Bharadiya, 2023). In this regard, Table 1 presents use cases linked to AI in the business environment.

5.2 *AI Applications in the Development of New Lines of Business*

Digital transformation has become an imperative for companies wishing to remain competitive in an ever-changing business environment (Weiss et al., 2016). However, this transformation goes beyond the adoption of new technologies; it involves a reassessment and restructuring of the organization with the goal of favouring sustainability in the future. The widespread integration of AI applications into the development trajectory of new enterprises is a paradigm shift in modern business endeavors. Thus, new models of entrepreneurship are emerging due to the increasing digitization and development of AI (Blanco-González-Tejero et al., 2023). Consequently, the seamless integration of AI into the fabric of business strategy is emerging as a key element in the search for pioneering entrepreneurial trajectories. In this way, AI can inspire the creation of disruptive business models (Brock & Von Wangenheim, 2019). Thus, when considering the generation of new models within the organization itself, the concept of intrapreneurship has always been highly explored (Hernández-Perlines et al., 2023). In this sense, digital transformation in intrapreneurship has been considered a key point in the development of companies in the real economy (Cheng et al., 2021). Thus, in business processes in maturity or decline, it will be essential to deepen and link digital strategies that enable the development of specific applications of AI in the context of the creation of new lines of business, thus giving rise to intrapreneurship.

Table 1 Examples of AI use cases in the enterprise world

Impact of AI on businesses	Use case
Advanced automation	Automation of customer service processes using chatbots with natural language processing (NLP) capabilities to resolve complex customer queries, reducing the burden on human agents.
Real-time data analytics	Real-time monitoring of the supply chain using sensors and real-time data analytics to detect and respond to logistical issues, such as delivery delays or product quality problems. This enables informed decisions to be made immediately.
Personalization	Video streaming platform that uses AI algorithms to analyze user viewing habits and deliver highly personalized content recommendations, increasing user retention and viewing time.
Advanced customer interaction	Virtual customer service assistant that uses AI to understand and effectively respond to customer queries, delivering a more satisfying customer service experience and improving customer loyalty.
Complex process optimization	Supply chain optimization using AI to predict demand, manage inventories, schedule production and shipment routes in a holistic manner, leading to greater efficiency and reduced costs.
Accurate forecasting	Use of machine learning models to predict product demand in retail, enabling more accurate inventory management and forecast-based decisions for improved business planning.
Advanced research and development	Discovery of new chemical compounds in the pharmaceutical industry by using AI to analyze large data sets and perform molecular simulations, accelerating the research and development of innovative drugs.
Error and risk reduction	Application of AI in the early detection of financial fraud by analyzing anomalous patterns in transactions, helping to reduce risk and minimize losses.
Transformation of traditional sectors	Implementation of telemedicine using AI for medical diagnosis and teleconsultation, revolutionizing healthcare by enabling wider and more efficient access to medical care, especially in remote areas.

6 The Future of AI in the Development of New Business Opportunities

The horizon of the business landscape is at the confluence of a transformation (T. H. Davenport & Ronanki, 2018) driven by the evolutionary trajectory of AI. This exposition delves into the prospective contours of AI's trajectory, illuminating its envisioned role in shaping the landscape of new business opportunities. By examining the symbiotic relationship between technological advances and business innovation, this analysis strives to unravel the potential paradigm shifts that AI induces, spanning diverse industries and market dimensions. As AI continues its growth, its latent capabilities for predictive analytics, autonomous decision-making, and enhanced human creativity portend a future replete with untapped business potentials, requiring businesses to navigate with discernment in this era of technological convergence and emerging economic models.

7 Implications and Vulnerabilities

This technology can optimize extremely complex and multifaceted business processes (Wirtz, 2011). The adoption and application of AI in the business environment brings with it significant opportunities (Ransome, 2010), but it also presents several challenges and considerations that companies and entrepreneurs must address strategically. Some of the key implications and challenges are detailed here:

1. **Automation and Role Transformation**: AI-driven automation may lead to the redefinition of roles and responsibilities in the workforce (Farrow, 2019; Papageorgiou et al., 2016). Some routine tasks could be replaced by AI systems, which will require reallocation of skills and focus on more creative and strategic tasks (Ng et al., 2021).
2. **Personalization and Customer Experience**: AI enables personalization (Wang et al., 2018) on a large scale, leading to higher expectations for tailored customer experiences. Companies must balance the convenience of personalization with data privacy and security.
3. **Data-Driven Decision Making**: AI provides data-informed decision making and deep analytics (Ng et al., 2021). However, this also raises the need to understand and assess the quality of the data used to avoid biased or incorrect decisions.
4. **Innovation and New Business Models**: AI can inspire the creation of new business models (Carneiro et al., 2017) and the disruption of traditional industries. Companies must be prepared to adapt to rapid changes in markets and competition.
5. **Ethics and Responsibility**: The debate has accelerated in recent times due to concerns about the use and impact of increasingly advanced algorithms, the increased availability of computational resources, as well as the increase in the number of data that can be used to analyze (Hall & Pesenti, 2017).

   Not knowing for sure what is carried out within AI is one of the main problems faced by ethicists. It is logical to think that for these experts to carry out their work they should be familiar with the technology they are trying to legislate on given that no one knows how the most advanced algorithms do what they do, which is a big problem (Knight, 2017). Despite this, it is necessary to reflect and discuss the ethical issues that may arise both in their creation and in their use (biases).
6. **Challenges**

   Investment and Resources: Successful implementation of AI may require significant investments in technology (Canhoto & Clear, 2020), talent and training. Companies, especially smaller ones, must balance these investments with the expected benefits.
7. **AI Talent Shortage**: Demand for AI experts exceeds the current supply in many cases. Companies struggle to attract and retain professionals with AI skills (Kaplan & Haenlein, 2020), which can hinder effective implementation.

8. **Data Privacy and Security**: Data collection and analysis can raise privacy (Martin & Murphy, 2017) and security concerns. Companies must comply with regulations and ensure the protection of sensitive data.
9. **Bias and Fairness**: AI models can incorporate biases (Villasenor, 2019) and discriminations present in training data. Identifying and mitigating these biases is a challenge to ensure fairness in decisions made by AI.
10. **Organizational Adaptation**: The adoption of AI may require a reconfiguration of processes and a cultural adaptation in the organization (Arias-Pérez & Cepeda-Cardona, 2022). Resistance to change (Choudhury et al., 2020) and the need for a new mindset may be challenges to overcome (Fountaine et al., 2019).
11. **Labor Relations**: Task automation may raise concerns among workers about job security (Buera & Kaboski, 2012). Companies must consider how to balance the adoption of AI with the well-being of their workforce (Huang & Rust, 2018).

8 Conclusions

In the current context of constant technological evolution and business dynamism, the role of AI in shaping new paradigms of business development acquires strategic and prospective relevance. This chapter has delved into the critical intersection between AI and business innovation, exploring its evolution and different examples of applications in the identification and materialization of emerging business opportunities. As AI takes hold as a catalyst for transformation (Arnold & Scheutz, 2018), it can be seen to not only redraw the frontier of commercial possibilities, but also incite the redefinition of strategic approaches and organizational cultures, anticipating a business future shaped by unprecedented synergies between technology and entrepreneurial innovation. By adopting an AI-informed strategy (Arnold & Scheutz, 2018), companies can explore emerging opportunities, innovate their business models, and offer unique value propositions that enable them to maintain a competitive advantage in a constantly evolving business environment. Therefore, this chapter provides insight into the business potential that AI brings, enabling business leaders and researchers to delve into each of the issues raised.

References

Aguirre, S., & Rodriguez, A. (2017). Automation of a business process using robotic process automation (RPA): A case study. *Communications in Computer and Information Science, 742*, 65. https://doi.org/10.1007/978-3-319-66963-2_7

Akerkar, R. (2019). *Artificial intelligence for business*. Springer.

Akter, S., Michael, K., Uddin, M. R., McCarthy, G., & Rahman, M. (2022). Transforming business using digital innovations: The application of AI, blockchain, cloud and data analytics. *Annals of Operations Research*, 1–33.

Albukhitan, S. (2020). Developing digital transformation strategy for manufacturing. *Procedia Computer Science, 170*, 664–671.
Anagnoste, S. (2017). Robotic automation process - the next major revolution in terms of back office operations improvement. *Proceedings of the International Conference on Business Excellence, 11*. https://doi.org/10.1515/picbe-2017-0072
Arias-Pérez, J., & Cepeda-Cardona, J. (2022). Knowledge management strategies and organizational improvisation: What changed after the emergence of technological turbulence caused by artificial intelligence? *Baltic Journal of Management, 17*(2), 250–265.
Bag, S., Gupta, S., Kumar, A., & Sivarajah, U. (2021). An integrated artificial intelligence framework for knowledge creation and B2B marketing rational decision making for improving firm performance. *Industrial Marketing Management, 92*, 178–189.
Bharadiya, J. P. (2023). Machine learning and AI in business intelligence: Trends and opportunities. *International Journal of Computer (IJC), 48*(1), 123–134.
Bishop, C. M., & Nasrabadi, N. M. (2006). *Pattern recognition and machine learning* (Vol. 4). Springer.
Blanco-González-Tejero, C., Ribeiro-Navarrete, B., Cano-Marin, E., & McDowell, W. C. (2023). A systematic literature review on the role of artificial intelligence in entrepreneurial activity. *International Journal on Semantic Web and Information Systems (IJSWIS), 19*(1), 1–16.
Bokrantz, J., Skoogh, A., Berlin, C., & Stahre, J. (2017). Maintenance in digitalised manufacturing: Delphi-based scenarios for 2030. *International Journal of Production Economics, 191*, 154–169.
Brock, J. K.-U., & Von Wangenheim, F. (2019). Demystifying AI: What digital transformation leaders can teach you about realistic artificial intelligence. *California Management Review, 61*(4), 110–134.
Buera, F. J., & Kaboski, J. P. (2012). The rise of the service economy. *American Economic Review, 102*(6), 2540–2569.
Caner, S., & Bhatti, F. (2020). A conceptual framework on defining businesses strategy for artificial intelligence. *Contemporary Management Research, 16*(3), 175–206.
Canhoto, A. I., & Clear, F. (2020). Artificial intelligence and machine learning as business tools: A framework for diagnosing value destruction potential. *Business Horizons, 63*(2), 183–193.
Cardholm, L. (2016). Demonstrating business value of security investments in the age of digitalization. *International Journal of Innovation in the Digital Economy (IJIDE), 7*(3), 1–25.
Carneiro, N., Figueira, G., & Costa, M. (2017). A data mining based system for credit-card fraud detection in e-tail. *Decision Support Systems, 95*, 91–101. https://doi.org/10.1016/j.dss.2017.01.002
Chauhan, S., Akhtar, A., & Gupta, A. (2022). Customer experience in digital banking: A review and future research directions. *International Journal of Quality and Service Sciences, 14*(2), 311–348.
Cheng, L., Varshney, K. R., & Liu, H. (2021). Socially responsible ai algorithms: Issues, purposes, and challenges. *Journal of Artificial Intelligence Research, 71*, 1137–1181.
Chi-Hsien, K., & Nagasawa, S. (2019). Applying machine learning to market analysis: Knowing your luxury consumer. *Journal of Management Analytics, 6*(4), 404–419.
Choudhury, P., Starr, E., & Agarwal, R. (2020). Machine learning and human capital complementarities: Experimental evidence on bias mitigation. *Strategic Management Journal, 41*(8), 1381–1411.
Codini, A. P., Abbate, T., & Petruzzelli, A. M. (2023). Business model innovation and exaptation: A new way of innovating in SMEs. *Technovation, 119*, 102548.
Cortez, R. M., & Johnston, W. J. (2017). The future of B2B marketing theory: A historical and prospective analysis. *Industrial Marketing Management, 66*, 90–102.
Da Xu, L., Lu, Y., & Li, L. (2021). Embedding blockchain technology into IoT for security: A survey. *IEEE Internet of Things Journal, 8*(13), 10452–10473.
Davenport, T., Guha, A., Grewal, D., & Bressgott, T. (2020). How artificial intelligence will change the future of marketing. *Journal of the Academy of Marketing Science, 48*, 24–42.

Davenport, T. H., & Ronanki, R. (2018). Artificial intelligence for the real world. *Harvard Business Review, 96*(1), 108–116.

Deng, L., & Yu, D. (2014). Deep learning: Methods and applications. *Foundations and Trends® in Signal Processing, 7*(3–4), 197–387.

Dwivedi, Y. K., Hughes, L., Ismagilova, E., Aarts, G., Coombs, C., Crick, T., Duan, Y., Dwivedi, R., Edwards, J., & Eirug, A. (2021). Artificial intelligence (AI): Multidisciplinary perspectives on emerging challenges, opportunities, and agenda for research, practice and policy. *International Journal of Information Management, 57*, 101994.

Enholm, I. M., Papagiannidis, E., Mikalef, P., & Krogstie, J. (2022). Artificial intelligence and business value: A literature review. *Information Systems Frontiers, 24*(5), 1709–1734.

Farrow, E. (2019). To augment human capacity—Artificial intelligence evolution through causal layered analysis. *Futures, 108*, 61–71.

Feng, C. M., Park, A., Pitt, L., Kietzmann, J., & Northey, G. (2021). Artificial intelligence in marketing: A bibliographic perspective. *Australasian Marketing Journal, 29*(3), 252–263.

Fink, O., Netland, T., & Feuerriegelc, S. (2021). Artificial intelligence across company borders. *Communications of the ACM, 65*(1), 34–36.

Fountaine, T., McCarthy, B., & Saleh, T. (2019). Building the AI-powered organization. *Harvard Business Review, 97*(4), 62–73.

Giesen, E., Riddleberger, E., Christner, R., & Bell, R. (2010). When and how to innovate your business model. *Strategy & Leadership, 38*(4), 17–26.

Grewal, D., Hulland, J., Kopalle, P. K., & Karahanna, E. (2020). The future of technology and marketing: A multidisciplinary perspective. *Journal of the Academy of Marketing Science, 48*, 1–8. Springer.

Gutiérrez, J. A., & Molina, B. (2016). Identificación de técnicas de minería de datos para apoyar la toma de decisiones en la solución de problemas empresariales. *Revista Ontare, 3*(2), 33–51.

Hall, W., & Pesenti, J. (2017). *Growing the artificial intelligence industry in the UK*.

Heaton, J. (2018). Ian goodfellow, Yoshua Bengio, and Aaron Courville: Deep learning: The MIT Press, 2016, 800 pp, ISBN: 0262035618. *Genetic Programming and Evolvable Machines, 19*(1–2), 305–307.

Hernández-Perlines, F., Araya-Castillo, L., Millán-Toledo, C., & Cisneros, M. A. I. (2023). Socioemotional wealth: A systematic literature review from a family business perspective. *European Research on Management and Business Economics, 29*(2), 100218.

Hofmann, P., Samp, C., & Urbach, N. (2020). Robotic process automation. *Electronic Markets*. https://doi.org/10.1007/s12525-019-00365-8

Huang, M.-H., & Rust, R. T. (2018). Artificial intelligence in service. *Journal of Service Research, 21*(2), 155–172.

Iqbal, S., Altaf, W., Aslam, M., Mahmood, W., & Khan, M. U. G. (2016). Application of intelligent agents in health-care. *Artificial Intelligence Review, 46*, 83–112.

Janiesch, C., Zschech, P., & Heinrich, K. (2021). Machine learning and deep learning. *Electronic Markets, 31*(3), 685–695.

Jasiulewicz-Kaczmarek, M., & Gola, A. (2019). Maintenance 4.0 technologies for sustainable manufacturing-an overview. *IFAC-PapersOnLine, 52*(10), 91–96.

Jiang, F., Jiang, Y., Zhi, H., Dong, Y., Li, H., Ma, S., Wang, Y., Dong, Q., Shen, H., & Wang, Y. (2017). Artificial intelligence in healthcare: Past, present and future. *Stroke and Vascular Neurology, 2*(4).

Kane, G. (2019). The technology fallacy: People are the real key to digital transformation. *Research-Technology Management, 62*(6), 44–49.

Kaplan, A., & Haenlein, M. (2019). Siri, Siri, in my hand: Who's the fairest in the land? On the interpretations, illustrations, and implications of artificial intelligence. *Business Horizons, 62*(1), 15–25.

Kaplan, A., & Haenlein, M. (2020). Rulers of the world, unite! The challenges and opportunities of artificial intelligence. *Business Horizons, 63*(1), 37–50.

Knight, W. (2017). The dark secret at the heart of AI. *MIT Technology Review, 120*(3), 54–65.

Leal-Rodríguez, A. L., Sanchís-Pedregosa, C., Moreno-Moreno, A. M., & Leal-Millán, A. G. (2023). Digitalization beyond technology: Proposing an explanatory and predictive model for digital culture in organizations. *Journal of Innovation & Knowledge, 8*(3), 100409.

Legner, C., Eymann, T., Hess, T., Matt, C., Böhmann, T., Drews, P., Mädche, A., Urbach, N., & Ahlemann, F. (2017). Digitalization: Opportunity and challenge for the business and information systems engineering community. *Business & Information Systems Engineering, 59*, 301–308.

Ma, L., & Sun, B. (2020). Machine learning and AI in marketing–connecting computing power to human insights. *International Journal of Research in Marketing, 37*(3), 481–504.

Martin, K. D., & Murphy, P. E. (2017). The role of data privacy in marketing. *Journal of the Academy of Marketing Science, 45*, 135–155.

Matt, C., Hess, T., & Benlian, A. (2015). Digital transformation strategies. *Business & Information Systems Engineering, 57*, 339–343.

Miller, S. (2018). AI: Augmentation, more so than automation. *Asian Management Insights, 5*(1), 1–20.

Mishra, A. N., & Pani, A. K. (2021). Business value appropriation roadmap for artificial intelligence. *VINE Journal of Information and Knowledge Management Systems, 51*(3), 353–368.

Monostori, L., Kádár, B., Bauernhansl, T., Kondoh, S., Kumara, S., Reinhart, G., Sauer, O., Schuh, G., Sihn, W., & Ueda, K. (2016). Cyber-physical systems in manufacturing. *CIRP Annals, 65*(2), 621–641.

Mughal, A. A. (2018). Artificial intelligence in information security: Exploring the advantages, challenges, and future directions. *Journal of Artificial Intelligence and Machine Learning in Management, 2*(1), 22–34.

Mughal, A. A. (2022). Building and securing the modern security operations Center (SOC). *International Journal of Business Intelligence and Big Data Analytics, 5*(1), 1–15.

Nadeem, A., Abedin, B., Cerpa, N., & Chew, E. (2018). Digital transformation & digital business strategy in electronic commerce-the role of organizational capabilities. *Journal of Theoretical and Applied Electronic Commerce Research, 13*(2), 1–8. Multidisciplinary Digital Publishing Institute.

Ng, K. K. H., Chen, C.-H., Lee, C. K. M., Jiao, J. R., & Yang, Z.-X. (2021). A systematic literature review on intelligent automation: Aligning concepts from theory, practice, and future perspectives. *Advanced Engineering Informatics, 47*, 101246.

Othman, F., Bahrin, M. A., & Azli, N. (2016). Industry 4.0: A review on industrial automation and robotic. *Jurnal Teknologi, 78*(6–13), 137–143.

Papageorgiou, E., Christou, C., Spanoudis, G., & Demetriou, A. (2016). Augmenting intelligence: Developmental limits to learning-based cognitive change. *Intelligence, 56*, 16–27.

Ransome, J. W. R. J. F. (2010). *Cloud computing implementation, management, and security*. CRC Press.

Reim, W., Åström, J., & Eriksson, O. (2020). Implementation of artificial intelligence (AI): A roadmap for business model innovation. *AI, 1*(2), 11.

Rust, R. T., & Huang, M.-H. (2014). The service revolution and the transformation of marketing science. *Marketing Science, 33*(2), 206–221.

Scarpi, D., Pizzi, G., & Matta, S. (2022). Digital technologies and privacy: State of the art and research directions. *Psychology & Marketing, 39*(9), 1687–1697.

Sharma, S. K., & Sharma, M. (2019). Examining the role of trust and quality dimensions in the actual usage of mobile banking services: An empirical investigation. *International Journal of Information Management, 44*, 65–75.

Shrestha, Y. R., Krishna, V., & von Krogh, G. (2021). Augmenting organizational decision-making with deep learning algorithms: Principles, promises, and challenges. *Journal of Business Research, 123*, 588–603.

Siderska, J. (2020). Robotic process automation-a driver of digital transformation? *Engineering Management in Production and Services, 12*(2), 21–31. https://doi.org/10.2478/emj-2020-0009

Silver, D., Hubert, T., Schrittwieser, J., Antonoglou, I., Lai, M., Guez, A., Lanctot, M., Sifre, L., Kumaran, D., & Graepel, T. (2018). A general reinforcement learning algorithm that masters chess, shogi, and go through self-play. *Science, 362*(6419), 1140–1144.

Sobczak, A. (2022). Robotic process automation as a digital transformation tool for increasing organizational resilience in polish enterprises. *Sustainability, 4*(3). https://doi.org/10.3390/su14031333

Valcozzena, S., Attanasio, G., Battistella, C., & Cantele, S. (2023). Exploring the sustainable business model of digital platforms. *New Business Models Conference Proceedings 2023*.

Villasenor, J. (2019). *Artificial intelligence and bias: Four key challenges*.

Wang, Y., & Kosinski, M. (2018). Deep neural networks are more accurate than humans at detecting sexual orientation from facial images. *Journal of Personality and Social Psychology, 114*(2), 246.

Wang, Y., Lu, X., & Tan, Y. (2018). Impact of product attributes on customer satisfaction: An analysis of online reviews for washing machines. *Electronic Commerce Research and Applications*. https://doi.org/10.1016/j.elerap.2018.03.003

Weiss, B. A., Pellegrino, J., Justiniano, M., & Raghunathan, A. (2016). Measurement science roadmap for prognostics and health management for smart manufacturing systems. *National Institute of Standards and Technology*, 100–102.

Wilson, H. J., & Daugherty, P. R. (2018). Collaborative intelligence: Humans and AI are joining forces. *Harvard Business Review, 96*(4), 114–123.

Wirtz, B. W. (2011). Business model management. *Design–Instrumente–Erfolgsfaktoren von Geschäftsmodellen, 2*(1).

Yarlagadda, R. T. (2018). The RPA and AI automation. *International Journal of Creative Research Thoughts (IJCRT)*. ISSN, 2320–2882.

Zhang, C., Xu, X., & Chen, H. (2020). Theoretical foundations and applications of cyber-physical systems: A literature review. *Library Hi Tech, 38*(1), 95–104.

Zhang, D. (2023). *Domain knowledge inspired deep learning: From general AI to specialized AI*. Rutgers The State University of New Jersey, Graduate School-Newark.

Zheng, P., Wang, H., Sang, Z., Zhong, R. Y., Liu, Y., Liu, C., Mubarok, K., Yu, S., & Xu, X. (2018). Smart manufacturing systems for industry 4.0: Conceptual framework, scenarios, and future perspectives. *Frontiers of Mechanical Engineering, 13*, 137–150.

Intelligent Transformation: Navigating the AI Revolution in Business and Technology

Ricardo Costa-Climent, Darek M. Haftor, and Marcin W. Staniewski

Abstract This chapter explores the revolutionary effects of artificial intelligence (AI) on modern business and technology. The chapter first presents a historical overview of AI development and varying definitions of AI, highlighting its central role in propelling business model innovation and technological progress. The multifaceted ethical challenges of AI are discussed, particularly within business contexts, in light of its potential to revolutionise the global economic landscape. A central idea is AI's transformative power to reshape industry norms and create novel economic possibilities. The discussion covers the dual function of AI (automation and augmentation of business operations), the ethical spectrum of AI application, and the implications for management and workforce dynamics. Emphasis is placed on the capacity of AI to enhance decision-making processes and its pivotal role in the development of strategies aligned with data-driven business models. The complex interplay between technical innovation and ethical considerations is examined. The chapter outlines a future where AI integration is transformative and responsible. By providing a thorough analysis of the technical opportunities and challenges of AI, the chapter offers a nuanced perspective of how AI continues to redefine the link between business and technology.

R. Costa-Climent (✉) · D. M. Haftor
Uppsala University, Uppsala, Sweden

University of Economics and Human Sciences, Warsaw, Poland
e-mail: ricardo.costacliment@im.uu.se; darek.haftor@im.uu.se

M. W. Staniewski
University of Economics and Human Sciences, Warsaw, Poland
e-mail: m.staniewski@vizja.pl

M. T. Del Val Núñez et al. (eds.), *Artificial Intelligence and Business Transformation*, Contributions to Management Science,
https://doi.org/10.1007/978-3-031-58704-7_2

1 Introduction

The advance of artificial intelligence (AI) is at the heart of performance improvements in technology. It has also driven the development of Industry 4.0. The technological breakthrough of Industry 4.0, which is attributed to AI, has facilitated human-machine interactions, reshaped the logic of business models, and transformed human lifestyle and quality of life. The integration of AI is creating a more intelligent and more innovative world. Examples such as route and traffic mapping in Google Maps, fare pricing in services such as Uber and Lyft, Facebook friend suggestions, email spam filters, online shopping recommendations, and cancer detection are just some of the AI-enabled technological innovations that simplify modern life. The rapid penetration of AI in various industries has prompted companies to join the competition and transform into AI-centric firms. In the early stages of the expansion of AI technologies, only a small sector with high purchasing power was able to take advantage of the growing opportunities of AI and engage in entrepreneurial activity (Apetrei et al., 2019). However, the use of AI-based technologies has fast become a driving force for businesses, strategists, visionaries, entrepreneurs, and researchers to devise new strategies and forge new sources of business value based on intelligent transformation.

Although AI is already present in many business sectors, the understanding of how it can be applied to create unique value is limited (Kemp, 2023). While some organisations successfully deploy AI technology to create unparalleled value, others fail. Quixey, Juicero, the Amazon Fire Phone, and Google Glass (Klein, 2022), are just some examples. Consequently, the following question arises: What conditions are required for successful value creation and appropriation from AI use?

To answer this question, this descriptive analytical paper explores the characteristics and capabilities of AI in the business context and its role in the intelligent transformation of business models. In exploring this issue, the chapter has two aims. First, it seeks to identify the role of AI in understanding business model innovation and transformation derived from related technologies. Second, it aims to show that the transformative potential of AI means that there is a need for theoretical development commensurate with its impact on society. This theoretical development must go beyond computational research or sociological accounts of human agency.

This chapter reviews the foundations of AI, its application in business models, and its role in intelligent transformation. Intelligent transformation is at the vanguard of a comprehensive revolution unfolding across the global business and technology landscape. Intelligent business transformation improves the quality of information dissemination, strengthens innovation capacity, and expands business boundaries, thereby reducing customer concentration (Mao et al., 2022). The role of AI in transforming business models is not just a trend. It is a profound shift that promises to redefine how industries operate and create value. On the verge of this AI revolution, it is crucial to realise that these changes are not incremental but pivotal, reshaping the nature of economic and technological paradigms.

The scientific literature shows that the transition to an intelligent business model system revolves around 10 technological innovations: big data analytics, the Internet of things, extended reality, additive manufacturing, cyber-physical systems, cloud services for products, integrated engineering systems, flexible manufacturing lines, simulation of virtual models, and cybersecurity (Atieh et al., 2023). The early sections of this chapter describe the underpinnings of this shift, underscoring not only the transformative potential of AI but also the urgency with which businesses must enter this new era.

The importance of intelligent transformation extends beyond technological advancement. It entails the emergence of new strategies, the disruption of long-standing industry practices, and the creation of unprecedented economic opportunities. As AI technologies mature, they drive innovation in business models, allowing for levels of efficiency, personalisation, and strategic foresight previously deemed unattainable.

Hence, this chapter considers the ethical challenges and issues arising from AI adoption. It lays the groundwork for a comprehensive review of the role of AI as a harbinger of change. In turn, it shows the need for more substantive theoretical and research support to broaden the understanding of this topic. By highlighting the transformative impact of AI from the outset, the chapter aims to give readers an understanding of the magnitude of the ongoing revolution and the many challenges and opportunities it creates for businesses worldwide.

2 Foundations of Artificial Intelligence

The inspiration for AI development was the desire to create autonomous machines with human-like thinking capabilities. AI emerged as an academic discipline in the 1950s. For half a century, it then became an area of obscurity in terms of practical and scientific interest (Haenlein & Kaplan, 2019). Certain aspects have a major impact on today's organisational world. These aspects include the complexity of the environment, which requires the search for new solutions, and the scarcity of resources, particularly the skills required to navigate the changes happening in business. Additionally, technological advancements are closely linked to the ever-growing trend of digitalisation (Troilo, 2023). Thanks to the rise of big data, improvements in computing power, and algorithms, AI has re-entered the business environment and people's daily lives. It has been applied in numerous ways in different sectors (Lee et al., 2019). Such applications involve understanding how machines can use language, create abstractions and concepts, and tackle complex problems.

The definition of AI varies depending on its goals and domain. However, a core feature of all definitions seems to be the ability to mimic human cognitive functions such as problem-solving and learning (Lee et al., 2019). For instance, Kaplan and Haenlein (2019) define AI as the ability of a system to interpret external data, learn from that data, and use that learning to achieve specific goals and tasks through flexible adaptation.

Although there are many AI applications, AI systems can be explained using a standard input-process-output model. AI information systems require structured data (in numerical form) or unstructured data (text, audio, images, etc.) from their environment (inputs). AI systems, which in essence are instances of machine learning (ML), manipulate that data to create value (processes). They then return the information (outputs) to the environment, either with human assistance or autonomously (Paschen et al., 2019).

2.1 Types of Artificial Intelligence

AI can be classified in terms of choice of approach, objective, scope, or discipline. For example, when classifying AI by stage of evolution, there is general AI and narrow AI. General AI systems exhibit behaviour that is indistinguishable from that of humans in all aspects (Haenlein & Kaplan, 2019). Narrow AI assists a human but is not autonomous. It lacks ethics, is standardised for its function, and cannot pass the Turing test (Etzioni & Etzioni, 2017).

Classifying AI by type of intelligence (cognitive, emotional, or social intelligence) leads to a classification of analytical AI, human-inspired AI, and humanised AI (Haenlein & Kaplan, 2019). Analytical AI only has characteristics consistent with cognitive intelligence. It is used for fraud detection in financial services, image recognition, and self-driving cars. Human-inspired AI has elements of both cognitive and emotional intelligence. Applications include recognising emotions during interactions with customers or job interviews. Humanised AI shows features of cognitive, emotional, and social intelligence. It refers to systems capable of self-awareness. It is not yet available (Haenlein & Kaplan, 2019).

It is also possible to differentiate between symbolic AI and neural AI depending on the degree of similarity to human cognitive capabilities derived from these systems. Symbolic AI is based on explicitly representing knowledge through symbols and logical rules. This approach uses a high-level structure to express concepts and relationships, enabling deductive reasoning. Symbolic AI systems manipulate symbols and apply logical rules to perform tasks such as natural language processing (NLP), decision making, and complex problem solving. This form of AI tends to be relatively transparent and understandable because the system's decisions can be traced back to the logical rules and symbols used to programme it. However, creating and maintaining symbolic systems can be resource intensive in terms of human effort because they require an explicit representation of knowledge and rules (Lee et al., 2019).

Neural AI is inspired by the functioning of the human brain. It uses artificial neural networks to perform specific tasks. These networks consist of layers of interconnected nodes, mimicking the structure of biological neurons. The strength of neural AI lies in its ability to learn patterns and adapt through training on large data sets. This approach is practical in tasks such as pattern recognition, computer vision, and high-volume unstructured data processing (Lee et al., 2019). Although decisions made by neural systems may be more difficult to interpret than those of symbolic systems, ML and adaptive capabilities are major strengths.

2.2 Machine Learning and Neural Networks

ML is a subfield of AI in computer science. According to Arthur Samuel (1959), it gives computers the ability to learn without being explicitly programmed (Ongsulee, 2017). AI inputs external data from the Internet of things (IoT) or other large data sources to identify underlying rules and patterns following some ML approach (Kaplan & Haenlein, 2019). ML encompasses computational procedures that allow AI systems to learn from experience and progressively improve their performance without being explicitly programmed by humans (Paschen et al., 2019; Russell & Norvig, 2016; Tecuci, 2012). The training system stores previous data, problem solving and reasoning processes, and results in its knowledge base (Paschen et al., 2020).

ML is not an augmentation of *general* AI that could lead machines to replace humans in all aspects of human cognition. It focuses on a particular aspect of intelligence: *prediction* (Agrawal et al., 2018). Statistical prediction can be defined as using existing data to fill in missing information. Prediction is valid because it is an input to decision making. Prediction is of no value if no decision is made. Hence, every prediction task perfectly complements a decision task (Agrawal et al., 2019).

There are three broad types of ML processes: supervised ML, unsupervised ML, and reinforcement ML (Haenlein & Kaplan, 2019; Paschen et al., 2020). With supervised ML, the AI system receives a training data set that contains the correct answers. The computer uses this training data to learn patterns and develop rules for future instances of the same problem.

With unsupervised ML, the computer uses training data that is not labelled with the correct answers. Its task is to identify patterns or relationships between data points (e.g. products purchased together). Finally, in reinforcement ML, an AI system learns from its own experience. It consists of finding a balance between exploring uncharted data and exploiting current knowledge based on experiences (Kaelbling et al., 1996).

The strategies of companies such as Google, Amazon, Apple, and Uber rely on the use of data. Their success shows the ability of data-processing-based technologies such as ML to create value and improve business performance. Amazon began as an online bookseller, fundamentally changing how customers shopped for books. eBay established itself as a leading online auctions platform. Google and Yahoo offered new ways for people to search for information and new opportunities for advertisers (Amit & Zott, 2020). ML is based on improving the price-performance ratio of computer processing technology, data storage, data management, and network technologies (Agrawal et al., 2018). In combination, these technologies make

AI essential to enable platforms, products, and services to generate user value (Gregory et al., 2021). While the literature on the effect of information technology (IT) on firm performance is vast, it offers little insight into the role of ML in this specific aspect.

Today, AI is under scrutiny because of advances in neural networks and deep learning technologies, which form the basis for most applications labelled as AI. For example, deep learning is the basis of the image recognition algorithms used by Facebook, the voice recognition algorithms that power smart speakers, and self-driving cars (Haenlein & Kaplan, 2019).

This advance in AI is essentially driven by two factors: the availability of big data and hardware accelerators such as graphics processing units, or GPUs, and tensor processing units, or TPUs (Abadi, 2016). Companies such as ORBCOMM, Microsoft, and Facebook are making this technology commercially available as deep learning libraries, personal and professional agents in the form of application programming interfaces (APIs), bots, chatbots, and other exciting products. They all increase business value by making products and services intelligent. This AI-driven paradigm shift is shaping other business contexts by improving employees' mechanical, analytical, and intuitive skills.

AI has reached a point where it can make financial decisions in the real world, chat with people, play against people, and work hand-in-hand with humans. Behind all these real-world applications is an AI-driven system or intelligent agent.

3 Artificial Intelligence as a Driver of Business Model Innovation

3.1 Business Model Innovation

A complete reconfiguration of a company's business is what business model innovation is all about (Zott & Amit, 2017). "A business model depicts the content, structure, and governance of transactions designed so as to create value through the exploitation of business opportunities" (Amit & Zott, 2001, p. 511). According to this definition, in the realm of e-business, there are four principal sources of value creation: efficiency, complementarities, lock-in, and novelty (Amit & Zott, 2001). Table 1 outlines the benefits and key factors of AI-driven business models that are critical for creating and sustaining competitive advantage and value creation.

Business model innovation is defined as a substantial change in operations that results in improved performance and value creation. AI has propelled this innovation across numerous sectors, including technology, media, consumer products, financial services, healthcare, manufacturing, energy, and the public sector. Introducing novel components in individual business models and creating a value architecture are key elements of this innovation, according to scholars such as Lee et al. (2019).

Table 1 AI-driven business models

Business model theme	Related AI-driven business model key points
Efficiency	Cost reduction, improved service quality, increased coordination and productivity, delivery efficiency
Lock-in	Development of algorithms, democratisation of AI, data-driven operations, scalable ecosystem integration
Novelty	Personalised products and services, new approaches to business models, outcome-based services
Complementarity	Real-time data analytics for personalised customer experiences, ethical AI ensuring fairness and transparency

However, adopting AI for business model innovation is not without its challenges. Identifying and implementing tailored digital innovations and understanding how to design, customise, evaluate, and market intangible offerings are prominent challenges. Some companies have leveraged digital technology to create significant value, while others have failed, with notable examples including Nokia, Blockbuster, and Kodak (Costa Climent & Haftor, 2021). These challenges are particularly evident in business models based on advanced digitally enabled services. In such cases, the offering is not a specific product or service but the promise of a given customer outcome (Parida et al., 2019).

As exemplified by Google, adopting AI for business model innovation is underpinned by a virtuous cycle. The availability of massive amounts of data enables organisations to create excellent products and services, which attract more users, thus generating more data. Data quality and quantity are essential to drive AI. Companies must focus on acquiring high-quality data and establishing a robust data infrastructure.

Implementing AI can transform business model innovation processes across an organisation, especially in industrial sectors. This change involves revisiting the underlying logic of how AI technology is included in value offerings to interact with work, organisational functions, and business processes to ensure value delivery, capture, and competitiveness (Iansiti & Lakhani, 2020).

The business model advantages of AI are diverse. They include cost reduction, improved service quality, increased coordination and productivity, and delivery efficiency. By expanding in scale and scope, AI-driven business models can find new ways of creating, delivering, and capturing value, thereby driving competitiveness in the marketplace (Sjödin & Palmié, 2021). AI-enabled companies are better positioned to expand their value space by using automated insights to develop data-driven work approaches and improve customer interactions (Sjödin & Palmié, 2021).

However, technological capabilities in AI are insufficient on their own. Firms must also grasp how to commercialise technology effectively by innovating appropriate AI business models. In this context, there is a need to reshape business model concepts to leverage disruptive technologies to the full. There is currently little theory on how the use of AI can positively affect business performance (Shollo et al., 2022). Åström et al. (2022), for example, highlighted the alignment of value

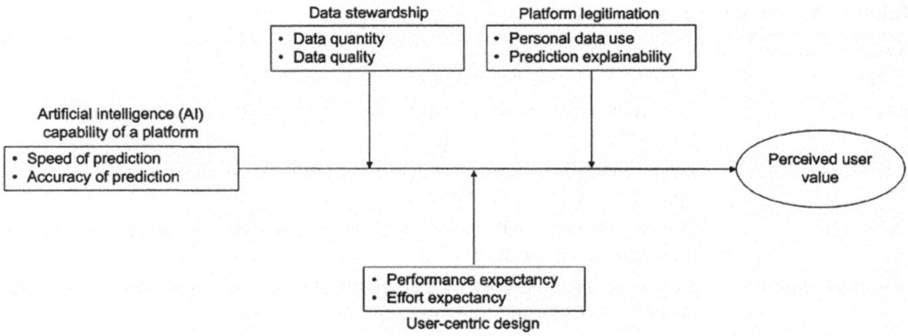

Fig. 1 Model of data network effects

creation and value capture mechanisms for AI providers in crafting successful AI business models. Their proposed framework involves defining the prerequisites for AI value creation, discovering effective value capture mechanisms, and formulating an offering for an AI business model. Drawing on data network effects theory, Gregory et al. (2021) argue that the value of a data network increases as more users contribute to and use the information in that network. The value perceived by users is related to the predictive ability and accuracy of AI (Gregory et al., 2021, 2022). Kemp (2023) recently developed the theory of situated AI, stressing the importance of organisational activities involved in the location of AI such as grounding, bounding, and recasting. Despite these theoretical advances in relation to the business value of AI use, the rapid development of emerging technologies demands further attention from scholars (Figs. 1 and 2).

3.2 *Transforming Business Models*

Digital transformation is radically changing the global business landscape. In this new reality, IoT technologies, intensive data sharing, and predictive analytics provide benefits such as process automation and optimisation. These benefits can improve productivity and profitability by saving costs, speeding up production, and reducing errors (Parida et al., 2019). The adoption and use of advanced technologies such as big data or AI at the right time and in the right way can help organisations increase business flexibility, improve customer service, and identify new business opportunities by processing information flows (Prokopenko et al., 2020). Despite the growing tendency to benefit from digital conversion, finding the right approach to use it in its true sense remains a challenge for many companies (Prokopenko et al., 2020).

Numerous companies have moved from a traditional business model to an e-business model. Amazon, Netflix, Alibaba, and eBay are online retail brands that have transformed the market through automation and AI processes. Their advanced

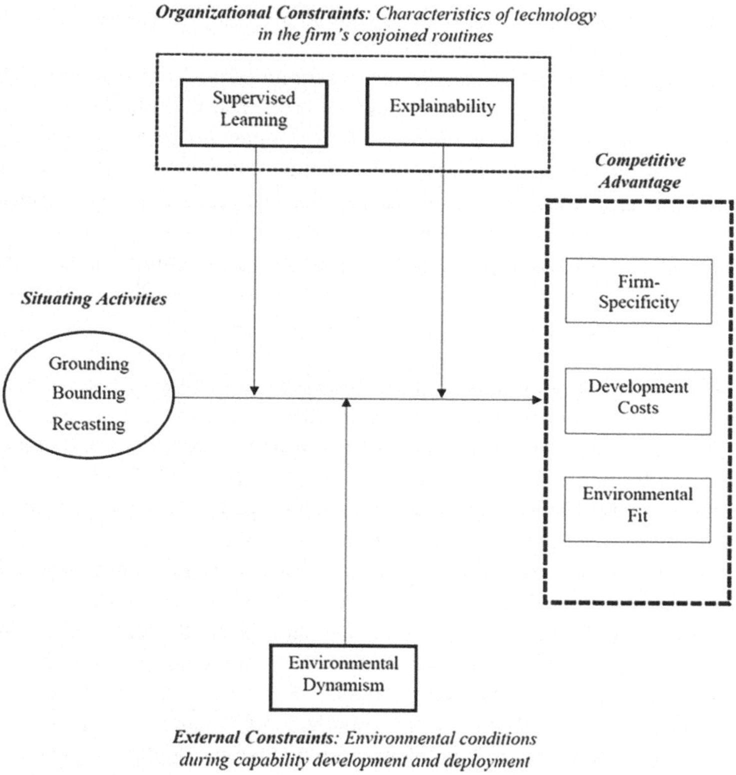

Fig. 2 The situated AI framework (Kemp, 2023)

technologies have created a better buying and selling experience for users through recommendation engines, sales prediction, warehouse automation, and innovative e-commerce platforms (Soni et al., 2019).

This process, as attractive as it may seem thanks to the overwhelming success of these examples, remains a challenge for entrepreneurs in any industry. Reasons for this challenge include the high costs of digital transformation and the fact that digitalisation is about more than just applying a collection of digital technologies. Instead, it involves a profound organisational change that not all organisations are prepared for, especially service or performance-based companies.

3.3 Servitisation

Digitalisation is a major challenge for industries rooted in traditional business models, especially those that trade in products or services. Adopting digital strategies is imperative for their survival in an increasingly competitive market. This digital shift affects traditional industries, particularly manufacturing, as explained by Ardolino et al. (2018) and Jovanović et al. (2021).

AI is becoming the most advanced form of digitalisation. Implementing AI goes hand-in-hand with the servitisation strategy of industrial manufacturers. In today's highly competitive job market, workers have to compete not only against each other but also against technological alternatives such as automation, ML, and AI. As a result, academic and practitioner evaluations of an individual's work-readiness have become increasingly important (Rasiah et al., 2019). Digital servitisation, as defined by Sjodin et al. (2021), involves transforming processes, capabilities, offerings, and business models to generate greater service value using digital technologies, such as AI.

AI offers vast possibilities for industrial manufacturers immersed in digital servitisation by providing added value and competitive advantage (Autio et al., 2018; Brock & Von Wangenheim, 2019; Nambisan, 2017). However, success is not guaranteed. Despite investment in digital infrastructure, technology, and data, many companies fail to increase value creation and capture from AI (Linde et al., 2020).

AI technology lays the foundation for success, but it is not enough on its own. New routines, skills, and processes are required so that firms can take full advantage of this technology and create customer value. Human interpretation and action are crucial because AI algorithms offer tentative solutions that need human validation and implementation (Tarafdar et al., 2019; von Krogh, 2018). B2B companies must undertake large-scale organisational transformation to develop AI capabilities and apply them effectively in their business model.

A commitment to developing interdependent AI capabilities, securing a data pipeline, developing algorithms, and democratising AI are fundamental steps. These capabilities are essential to identify, evaluate, and exploit AI-related business opportunities. In addition, business model transformation must be based on agile customer co-creation, data-driven delivery operations, and scalable ecosystem integration (Sjödin & Palmie, 2021). Developing a comprehensive strategy encompassing capabilities, business model transformation, and core principles is necessary.

3.4 Fully Benefiting from Digitalisation and AI in Business Model Innovation

Davenport et al. (2020) cite two separate applications of AI. First, AI can be used to analyse and interpret external data, which enables flexible learning and adaptation. Second, AI can be deployed to streamline business operations, extract meaningful information from data sets, and enhance interactions with employees and customers.

3.5 Personalisation and Customer Experience

AI has two key characteristics: self-learning and connectivity (Huang & Rust, 2022). These features make AI capable of adapting to changing customer needs. AI can automatically self-perfect itself by learning from inputs (e.g. big data and ML). ML refers to using computer algorithms (statistical and mathematical optimisation) to enhance an offering, and data analysis enables these algorithms to create precise predictions (Haftor et al., 2023). AI connectivity is most evident in IoT applications, which link machines, objects, and data streams to aid learning.

Companies employing AI use customers' past behaviour (such as previous purchases, clicks, selections, or ratings), current search keywords, and many other parameters to deliver personalised digital ads, coupons, deal suggestions, and emails or messages inviting customers to purchase specially chosen products (Soni et al., 2019). These recommendation systems build customer loyalty and can secure more sales by attracting valuable customers who visit the site more often and spend more.

ML thus represents an augmentation of a particular aspect of intelligence: *prediction* (Agrawal et al., 2018, 2019). Agrawal et al. (2019) define forecasting as the statistical approach of using existing data to fill in missing information. Forecasting is crucial in business management because it complements decision making.

Data volume, variety, and veracity are the basis of predictive models that allow data-driven products and services to be continuously improved through feedback from users who share personal data in real time (Gregory et al., 2021). The more users interact with products or services, the more the algorithms learn. Predictive patterns then improve, and better and more personalised products and services can be offered. This loop retains users and even brings new participants into the network. Positive relationships between network size and user-perceived value creation are known as data network effects (Gregory et al., 2021).

3.6 Business Process Automation

Organisations have long used AI-based solutions to automate routine operations and logistics tasks. Advances in computing power, exponential data growth, and new ML techniques allow organisations to use AI-based solutions for management tasks (Brynjolfsson & Mcafee, 2017).

Raisch and Krakowski (2021) describe the interdependent relationship between two broad applications of AI in organisations: automation and augmentation. Automation means that machines take over a human task, whereas augmentation means that humans and machines work closely together to perform a task. These dual applications of AI are interdependent in time and space, creating a paradoxical tension. Opting exclusively for augmentation or automation can fuel cycles of reinforcement that not only harm an organisation's performance but also have negative social implications. However, organisations that opt for both automation and augmentation can cope with this tension and even achieve complementarities that benefit business and society. ML plays a leading role in augmentation as an application of AI in organisations.

3.7 New Revenue Models

AI-driven personalisation is becoming a transformative element of innovative business model creation. The ability to collect, analyse, and interpret large amounts of data quickly and accurately allows companies to offer highly personalised products and services tailored to individual user needs. This customer-centric approach not only improves customer satisfaction but also creates opportunities for new approaches to business models.

The value creation of these new business models stems from the unique self-learning and predictive capabilities that characterise ML technologies. This value creation provided by ML depends on the speed and accuracy of the predictions it can make. Those attributes are in turn moderated by the quantity and quality of data, expectations of performance and effort, and legitimacy of the technology (Gregory et al., 2021). In light of these moderators of ML-based value creation, some key aspects of AI-based business models can be identified as leading to greater value perceived by users.

The first is deep customer insight. AI allows companies to analyse real-time data, from purchasing preferences to online behaviours. This data analysis can provide a deep understanding of each and every customer. This understanding goes beyond simple demographic segmentations. Instead, it enables precise personalisation based on individual habits and preferences. Another key aspect of AI-based business models is the outcome-based services approach. The ability to predict behaviours and outcomes through predictive algorithms leads to the creation of outcome-based services. Companies not only sell products but also offer solutions that deliver specific outcomes such as improved health, educational performance, and efficiency at home. Another aspect is continuous adaptation. AI-driven personalisation is continuous, not static. Business models can adapt and evolve with changing customer preferences and needs. This continuous adaptation improves customer loyalty and opens the door to improved subscription models and constantly evolving services. Finally, through internal process optimisation, AI can improve supply chains, inventory management, and other operational processes, ensuring businesses can efficiently meet customer expectations.

The increase in perceived user or customer value that companies can achieve using AI technologies is undeniable. However, given that the party that creates the value may not be able to capture it, value creation and value capture (or appropriation) should be considered distinct but interconnected processes (Lepak et al., 2007). For Moran and Ghoshal (1999), firms' strategies must address both value creation and value appropriation. This balance is at the heart of an effective business strategy.

Digital business strategies that base their value creation and value capture strategy on data, AI, and data learning systems rely heavily on IT to coordinate the actors involved (Bharadwaj et al., 2013). These business strategies are known as digital platforms or platform ecosystems (Lan et al., 2019). Platform ecosystems are networks of loosely coupled firms united through a shared technological infrastructure. This infrastructure provides the building blocks upon which participating firms can develop complementary products and services (Gawer, 2009, 2014). It offers a new locus of coordination that brings together participants to create value jointly (Ceccagnoli et al., 2012; Gawer & Henderson, 2007; Tiwana, 2015; West, 2003).

In the context of emerging technologies, a thoughtful and strategic design of the business model can complement other mechanisms, such as patenting or legal action, to profit from innovation (Snihur et al., 2021). The traditional approach to value creation and capture has shifted considerably. Before, firms possessed inimitable resources. Now, value co-creation occurs in interfirm networks or ecosystems, as noted by Schreieck et al. (2017). This shift involves capturing value from the market and appropriating value from within a network of partners.

Scalable ecosystem integration is central to efficient value capture. It stimulates collaboration between business units, customers, and ecosystem actors. The importance of ecosystems as catalysts for scalability, especially in AI-based digital servitisation strategies, is growing. To ensure the scalability of AI, value-enabling partnerships are critical. Creating dedicated partnerships to provide access to back-end AI competencies is essential for a business's continued development and scalability. Notable examples include integrating technology leaders such as Microsoft and IBM into the backbone of the business model by providing a core cloud infrastructure and algorithm support (Sjödin & Palmié, 2021).

A second crucial area for value generation is leveraging value-expansion partnerships involving small and medium-sized enterprises (SMEs), start-ups, and original equipment manufacturers (OEMs). These partners provide vital specialised expertise that augments internal AI capabilities and catalyses business model innovation and organisational transformation.

Finally, once an AI algorithm or data pipeline is developed, it becomes scalable with low marginal costs and network effects. The more users connect to an algorithm, the more powerful its learning base becomes. Hence, scaling AI across broader ecosystems and customers is important. It allows external actors to create new value propositions, generating revenue without a corresponding cost increase. There is potential for the reconfiguration of AI solutions through the use of applications from multiple ecosystem actors. Collaboration and scalability in digital ecosystems are essential so that AI-generated value can be captured effectively.

4 Developments in Technology and AI

Effective interaction and communication are crucial in today's business landscape. AI plays a prominent role in this area. New AI-derived applications and tools are emerging, and these tools will transform business dynamics (Regona et al., 2022). These innovations are redefining the way businesses approach communication, interactions, and decision making. One prominent example is ChatGPT (Deng & Lin, 2022). This tool can improve business communication by enabling more intuitive and efficient interactions with customers and employees. Its ability to understand context and generate consistent responses has led to practical applications in customer service, virtual assistance, and internal collaboration (Berg et al., 2023).

Advances in natural language processing (NLP) have contributed substantially to this improvement (Chowdhary & Chowdhary, 2020). Advanced NLP has permitted a deeper understanding of human language and can adapt to different contexts. Computer vision and image recognition have revolutionised the interpretation of images in real-time. These technologies have been used in fraud detection, quality control, and product customisation. Similarly, the automatic generation of multimedia content using AI has simplified the creation of images and videos. This feature has transformed marketing and advertising strategies, democratising the competition. In parallel, the combination of AI and robotics is driving the automation of physical tasks in the logistics, manufacturing, and inventory management sectors.

The integration of AI with blockchain technology marks a fundamental shift. While the AI applications discussed so far focus on improving processes and communication, blockchain disrupts an essential social structure: the financial system. Contract automation uses intelligent contracts, supported by the combination of AI and blockchain. It adds efficiency and transparency to business transactions. By leveraging blockchain technology, which works as an open, broadly distributed ledger, this innovation eliminates the need for banking intermediaries, providing a transparent and accessible record of all transactions. This change is not limited to operational improvement. It also reshapes the basis of financial interactions by decentralising and democratising the process.

Finally, advances in quantum computing (Khan & La Torre, 2021) present a promising opportunity for solving complex problems. Quantum computing has transformative potential by optimising operations and decision-making processes. These advances highlight the rapid evolution of AI and its major impact on efficiency and innovation in the business environment.

5 Ethical Challenges and Considerations

AI has achieved greater levels of accuracy than humans in various tasks, including facial recognition, language translation, and strategic games such as Go and chess. This progress has created opportunities for innovation in processes and products.

However, ethical concerns and challenges arise when integrating AI into business models.

The superior accuracy of intelligent agents in critical tasks such as breast cancer detection and language transcription raises big ethical questions. Bias in automated decisions stemming from biases in the data sets used to train models is a common concern. The ethical responsibility of companies that adopt AI is to ensure fairness and justice for all segments of society.

The inherent opacity of some AI models such as deep neural networks presents challenges of transparency and explainability. A lack of clarity in algorithms can create distrust among users and stakeholders. To build consumer trust and meet regulatory requirements, companies must be able to explain how decisions are made in an understandable way.

AI researchers and practitioners face the task of addressing ethical issues such as bias, trust, privacy, and security. These ethical challenges require deep reflection. Specific steps must be designed to ensure that the deployment of AI in business models is ethical, fair, and transparent. Only then can the benefits be maximised and the risks minimised.

5.1 Impact of Artificial Intelligence on Human Resources

Automation and AI are transforming the work environment, posing challenges for human resource departments. Implementing advanced technologies leads to the automation of routine tasks and repetitive assignments. This automation affects traditional manufacturing and administrative jobs.

Accordingly, companies must offer transition strategies for employees whose roles are at risk of automation. The reconfiguration of roles implies that workers will need to acquire new skills. The rapid evolution of technology means that the skills in demand in the labour market are changing rapidly. Companies must therefore invest in training and development programmes to ensure that their workforce is equipped with the skills needed to work effectively with emerging technologies.

Although automation may eliminate some roles, there are also new opportunities in creative, strategic, and decision-making areas. Businesses should redefine roles to take advantage of human strengths such as creativity, ethical judgement, and complex problem-solving. This approach entails a focus on creating more creative and strategic jobs, where collaboration between humans and technology can boost job performance and foster innovation. Proactively adapting to these changes in the employment landscape will ensure long-term sustainability and success for companies and their employees.

5.2 Data Privacy and Security Barriers

The massive collection of data used to train AI models has led to huge advances. However, it also raises ethical challenges regarding privacy. Informed consent, anonymisation, and algorithmic discrimination are important issues. The absence of explicit consent, a lack of anonymisation, and biases in training data can lead to ethical and social problems.

Data security is essential to mitigate privacy violations. Strategies such as robust encryption, access management and control, regular audits, and transparency in data practices are crucial. Implementing ethical data management policies helps build user trust and ensures that innovation in AI is developed ethically and sustainably. Addressing data privacy and security barriers involves overcoming ethical challenges and implementing advanced protection measures. Balancing innovation in AI with privacy protection is critical to create an environment where technology can thrive ethically and sustainably.

5.3 Bias and Fairness in Algorithms

Inherent biases in training data for AI algorithms can lead to discriminatory results, amplifying societal inequalities. These biases, related to gender, race, age, and other factors, appear in a variety of ways, contributing to discrimination by association, data representation bias, and contextual bias. For example, discrimination by association occurs when algorithms learn patterns based on biased historical data, replicating inequalities in future decisions. A lack of equal representation in the data can lead to inaccurate or biased predictions for certain groups, while the context of data collection can introduce biases based on systemic inequalities.

Specific strategies are required to mitigate these biases and ensure fairness in the implementation of algorithms. These strategies may include correcting biases in data sets, implementing fair learning algorithms, and carefully considering the data collection context. By addressing biases from the development phase, progress can be made towards more equitable and fair algorithms.

5.4 Responsibility and Accountability

The rapid expansion of AI has underscored the urgent need to establish clear responsibilities for AI development and application. The autonomy of AI systems and the opacity of some models are factors that hinder the understanding of decisions. This situation emphasises the importance of addressing responsibility from various perspectives.

The need for clear accountability stems from autonomy in decision making, the opacity of some models, and the ethical and social impacts of AI-based decisions. A lack of accountability could result in discriminatory or unfair outcomes, highlighting the importance of addressing risks and consequences.

Various strategies have been proposed to establish responsibilities and ensure accountability. Creating legal and regulatory frameworks, including transparency requirements and ethical standards, is essential. Transparency in development, the documentation of decisions, and processes makes external reviews easier. Ethical evaluation refers to including ethical and social considerations in the design of AI systems. Organisational accountability involves the designation of ethics officers and internal review processes. Stakeholder participation, where diverse stakeholders are involved in the development and decision-making process, ensures a variety of perspectives and reduces the risk of inadvertent bias. These strategies, when combined, provide a comprehensive approach to addressing responsibility and accountability in the development and use of AI systems.

5.5 Transparency and Explainability of Algorithms

The transparency and explainability of algorithms are fundamental aspects of the development and application of AI, especially in critical situations that directly affect people. Understanding how algorithms make decisions inspires trust, addresses ethical issues, and makes it possible to identify and correct potential biases. However, there are challenges in explaining AI models. The inherent complexity of many models, especially those based on deep learning, and the opacity of those considered black boxes are features that make it difficult to explain how they reach their decisions. Furthermore, finding a balance between accuracy and explainability presents a dilemma, especially in contexts where explainability is required.

Strategies have been proposed to increase transparency. Adequate documentation of processes to detail how data is collected and how models are trained can provide a clear understanding. Using interpretable rather than black box models makes explanations easier because it enables the use of visualisations and interactive tools. In addition, educating users and stakeholders on the basic principles of AI contributes to a broader understanding of AI and thus greater acceptance and trust.

5.6 Environmental Impact of Artificial Intelligence

The increasing adoption of AI has raised concerns about its environmental impact. The carbon footprint associated with training and implementing AI models has been calculated. Training models such as deep neural networks is complex and computationally and energy intensive. Implementing these models in hardware, especially in server and data centre infrastructures, can contribute substantially to the carbon footprint of AI.

To address these issues, some initiatives have been implemented to develop AI more sustainably and eco-efficiently. Optimising algorithms to reduce resource requirements, developing specialised hardware to increase energy efficiency, and using renewable energy are notable strategies. Federated learning, hardware recycling, and industry engagement are crucial when searching for sustainable solutions.

5.7 Legal Aspects and Regulatory Framework

The rapid evolution of AI has raised legal challenges that demand effective responses to reconcile technological innovation with the protection of individual rights and public safety. Challenges include legal liability in automated decisions, data protection and privacy, algorithmic discrimination and bias, and the need for transparency and explainability in AI models.

To address these challenges, up-to-date, flexible regulatory frameworks are needed to adapt to the constant evolution of technology. These frameworks must be agile enough to adjust to technological advances. They must also enable international collaboration to establish common standards. Finally, ethics and technology experts should be involved in writing these regulations.

Different regions are responding to these challenges by introducing specific legislation and regulations. Notable examples include the General Data Protection Regulation (GDPR) in the European Union, the Artificial Intelligence Innovation and Competition Act (ICA) in France, the California Consumer Privacy Act (CCPA) in the United States, and the Ethical Governance of AI Act in Canada. These examples illustrate the range of approaches to addressing the legal challenges of AI in different parts of the world. They reflect the ongoing effort to balance technological innovation with the protection of fundamental rights.

6 Conclusions

This descriptive analysis addresses the main aim of this chapter, namely to demonstrate the need for intensive theoretical development to support the adoption of AI in the business environment. AI can transform the global economy through technological innovations, scientific breakthroughs, and business activity. The steady rise of automation and AI over the last decade can predominantly be attributed to the increasing availability of big data and hardware accelerators such as GPUs and TPUs. These factors make AI the core technology driving automation and extreme connectivity, thus ushering in the fourth industrial revolution. The exceptional capabilities of intelligent agents in gaming, recognition, and classification offer opportunities for process and product innovation, such as assistive technologies for the disabled and elderly, while driving advances in the toy and gaming industry. Although this technological autonomy enhances life in many ways, it also raises security, privacy, and ethical challenges that must be carefully addressed, not only in practice but also from a multi-theoretical scientific perspective. AI-based technologies influence and are influenced by a range of theoretical fields, including information systems, computer science, economics, psychology, and medicine. Present and future research on AI should seek a meeting point between these disciplines to give a holistic understanding of the current and future role of AI in society in its present form.

The adoption of intelligent transformation strategies by corporations (both large companies and start-ups) to become AI users can strengthen the innovation process and global competitiveness. The pursuit of growth by adopting advanced AI technology drives intense technological competition and therefore increases investment in AI. The rise of investment in automation and AI in recent years implies that it has the potential to change the economy. Nevertheless, AI technology remains limited to some areas of the world, creating an AI gap. The uneven spread of this technology deepens economic, cultural, and social inequalities, creating a divide similar to the digital divide. This disparity, identified as AI's dark side (Soni et al., 2019), raises concerns about global inequality and the need to address it ethically and equitably. This issue should be analysed in depth to help public decision makers develop policies for social and economic equality.

The analysis in this paper is not simply a proposal idealising the application of AI in business. Likewise, it is not a pessimistic account of the conceptual and knowledge barriers that prevent a full understanding of the underlying implications of AI use. The motivation of this synthesis and analysis of the transformative role of AI is to show one way for business and economics to achieve legitimacy in the AI discourse.

References

Abadi, M. (2016, September). TensorFlow: Learning functions at scale. In *Proceedings of the 21st ACM SIGPLAN international conference on functional programming*, p 1.

Agrawal, A., Gans, J. S., & Goldfarb, A. (2018, May). Human judgment and AI pricing. In *AEA Papers and Proceedings 108*, 58–63. 2014 Broadway, Suite 305, Nashville, TN 37203: American Economic Association.

Agrawal, A., Gans, J. S., & Goldfarb, A. (2019). Artificial intelligence: The ambiguous labor market impact of automating prediction. *The Journal of Economic Perspectives, 33*(2), 31–50.

Amit, R., & Zott, C. (2001). Value creation in e-business. *Strategic Management Journal, 22*(6–7), 493–520.

Amit, R., & Zott, C. (2020). *Business model innovation strategy: Transformational concepts and tools for entrepreneurial leaders*. John Wiley & Sons.

Apetrei, A., Sánchez-García, J. L., & Sapena, J. (2019). The controversial link between entrepreneurial activity and inequality. *The International Entrepreneurship and Management Journal, 15*, 485–502.

Ardolino, M., Rapaccini, M., Saccani, N., Gaiardelli, P., Crespi, G., & Ruggeri, C. (2018). The role of digital technologies for the service transformation of industrial companies. *International Journal of Production Research, 56*(6), 2116–2132.

Åström, J., Reim, W., & Parida, V. (2022). Value creation and value capture for AI business model innovation: A three-phase process framework. *Review of Managerial Science, 16*(7), 2111–2133.

Atieh, A. M., Cooke, K. O., & Osiyevskyy, O. (2023). The role of intelligent manufacturing systems in the implementation of industry 4.0 by small and medium enterprises in developing countries. *Engineering Reports, 5*(3), e12578.

Autio, E., Nambisan, S., Thomas, L. D., & Wright, M. (2018). Digital affordances, spatial affordances, and the genesis of entrepreneurial ecosystems. *Strategic Entrepreneurship Journal, 12*(1), 72–95.

Berg, J., Raj, M., & Seamans, R. (2023). Capturing value from artificial intelligence. *Academy of Management Discoveries, 9*(4).

Bharadwaj, A., El Sawy, O. A., Pavlou, P. A., & Venkatraman, N. V. (2013). Digital business strategy: Toward a next generation of insights. *MIS Quarterly*, 471–482.

Brock, J. K. U., & Von Wangenheim, F. (2019). Demystifying AI: What digital transformation leaders can teach you about realistic artificial intelligence. *California Management Review, 61*(4), 110–134.

Brynjolfsson, E., & Mcafee, A. (2017). Artificial intelligence, for real. *Harvard Business Review, 1*, 1–31.

Ceccagnoli, M., Forman, C., Huang, P., & Wu, D. J. (2012). Cocreation of value in a platform ecosystem! The case of enterprise software. *MIS Quarterly*, 263–290.

Chowdhary, K., & Chowdhary, K. R. (2020). *Natural language processing* (pp. 603–649). Fund Artif Intell.

Costa Climent, C., & Haftor, D. M. (2021). Business model theory-based prediction of digital technology use: An empirical assessment. *Technological Forecasting and Social Change, 173*, 121174.

Davenport, T., Guha, A., Grewal, D., & Bressgott, T. (2020). How artificial intelligence will change the future of marketing. *Journal of the Academy of Marketing Science, 48*, 24–42.

Deng, J., & Lin, Y. (2022). The benefits and challenges of ChatGPT: An overview. *Frontiers in Computing and Intelligent Systems, 2*(2), 81–83.

Etzioni, A., & Etzioni, O. (2017). Incorporating ethics into artificial intelligence. *The Journal of Ethics, 21*, 403–418.

Gawer, A. (2009). *Platforms, markets and innovation: An introduction*. Edward Elgar.

Gawer, A. (2014). Bridging differing perspectives on technological platforms: Toward an integrative framework. *Research Policy, 43*(7), 1239–1249.

Gawer, A., & Henderson, R. (2007). Platform owner entry and innovation in complementary markets: Evidence from Intel. *Journal of Economics and Management Strategy, 16*(1), 1–34.

Gregory, R. W., Henfridsson, O., Kaganer, E., & Kyriakou, H. (2021). The role of artificial intelligence and data network effects for creating user value. *Academy of management review, 46*(3), 534–551.

Gregory, R. W., Henfridsson, O., Kaganer, E., & Kyriakou, H. (2022). Data network effects: Key conditions, shared data, and the data value duality. *The Academy of Management Review, 47*, 189–192.

Haenlein, M., & Kaplan, A. (2019). A brief history of artificial intelligence: On the past, present, and future of artificial intelligence. *California Management Review, 61*(4), 5–14.

Haftor, D. M., Costa-Climent, R., & Navarrete, S. R. (2023). A pathway to bypassing market entry barriers from data network effects: A case study of a start-up's use of machine learning. *Journal of Business Research, 168*, 114244.

Huang, M. H., & Rust, R. T. (2022). A framework for collaborative artificial intelligence in marketing. *Journal of Retailing, 98*(2), 209–223.

Iansiti, M., & Lakhani, K. R. (2020). *Competing in the age of AI: Strategy and leadership when algorithms and networks run the world*. Harvard Business Press.

Jovanović, J., Saqr, M., Joksimović, S., & Gašević, D. (2021). Students matter the most in learning analytics: The effects of internal and instructional conditions in predicting academic success. *Computers in Education, 172*, 104251.

Kaelbling, L. P., Littman, M. L., & Moore, A. W. (1996). Reinforcement learning: A survey. *Journal of Artificial Intelligence Research, 4*, 237–285.

Kaplan, A., & Haenlein, M. (2019). Siri, Siri, in my hand: Who's the fairest in the land? On the interpretations, illustrations, and implications of artificial intelligence. *Business Horizons, 62*(1), 15–25.

Kemp, A. (2023). Competitive advantages through artificial intelligence: Toward a theory of situated AI. *The Academy of Management Review*. (in press).

Khan, F. S., & La Torre, D. (2021). Quantum information technology and innovation: A brief history, current state and future perspectives for business and management. *Technology Analysis & Strategic Management, 33*(11), 1281–1289.

Klein, J. (2022). *Competing in the age of AI*. By Marco Iansiti and Karim Lakhani. Harvard Business Review Press.

Lan, S., Liu, K., & Dong, Y. (2019). Dancing with wolves: How value creation and value capture dynamics affect complementor participation in industry platforms. *Industry and Innovation, 26*(8), 943–963.

Lee, J., Suh, T., Roy, D., & Baucus, M. (2019). Emerging technology and business model innovation: The case of artificial intelligence. *Journal of Open Innovation: Technology, Market, and Complexity, 5*(3), 44.

Lepak, D. P., Smith, K. G., & Taylor, M. S. (2007). Value creation and value capture: A multilevel perspective. *The Academy of Management Review, 32*(1), 180–194.

Linde, L., Sjödin, D., Parida, V., & Gebauer, H. (2020). Evaluation of digital business model opportunities: A framework for avoiding digitalization traps. *Research-Technology Management, 64*(1), 43–53.

Mao, J., Zhao, Y., Yang, S., Li, R. Y. M., & Abbas, J. (2022). Intelligent transformation and customer concentration. *Journal of Organizational and End User Computing, 35*(2), 1–15.

Moran, P., & Ghoshal, S. (1999). Markets, firms, and the process of economic development. *The Academy of Management Review, 24*(3), 390–412.

Nambisan, S. (2017). Digital entrepreneurship: Toward a digital technology perspective of entrepreneurship. *Entrepreneurship Theory and Practice, 41*(6), 1029–1055.

Ongsulee, P. (2017, November). Artificial intelligence, machine learning and deep learning. In *2017 15th international conference on ICT and knowledge engineering (ICT&KE)*, pp. 1–6, IEEE.

Parida, V., Sjödin, D., & Reim, W. (2019). Reviewing literature on digitalization, business model innovation, and sustainable industry: Past achievements and future promises. *Sustainability, 11*(2), 391.

Paschen, J., Kietzmann, J., & Kietzmann, T. C. (2019). Artificial intelligence (AI) and its implications for market knowledge in B2B marketing. *The Journal of Business and Industrial Marketing, 34*(7), 1410–1419.

Paschen, U., Pitt, C., & Kietzmann, J. (2020). Artificial intelligence: Building blocks and an innovation typology. *Business Horizons, 63*(2), 147–155.

Prokopenko, O., Shmorgun, L., Kushniruk, V., Prokopenko, M., Slatvinska, M., & Huliaieva, L. (2020). Business process efficiency in a digital economy. *International Journal of Management, 11*(3), 122–132.

Raisch, S., & Krakowski, S. (2021). Artificial intelligence and management: The automation–augmentation paradox. *The Academy of Management Review, 46*(1), 192–210.

Rasiah, R., Turner, J. J., & Ho, Y. F. (2019). The impact of emotional intelligence on work performance: Perceptions and reflections from academics in Malaysian higher education. *Contemporary Economics, 13*(3), 269.

Regona, M., Yigitcanlar, T., Xia, B., & Li, R. Y. M. (2022). Opportunities and adoption challenges of AI in the construction industry: A PRISMA review. *Journal of Open Innovation: Technology, Market, and Complexity, 8*(1), 45.

Russell, S. J., & Norvig, P. (2016). *Artificial intelligence: A modern approach*. Pearson.

Samuel, A. L. (1959). Some studies in machine learning using the game of checkers. *IBM Journal of Research and Development, 3*(3), 210–229.

Schreieck, M., Wiesche, M., Krcmar, H. (2017, December). *The platform owner's challenge to capture value-insights from a business-to-business IT platform*. ICIS.

Shollo, A., Hopf, K., Thiess, T., & Müller, O. (2022). Shifting ML value creation mechanisms: A process model of ML value creation. *The Journal of Strategic Information Systems, 31*(3), 101734.

Sjödin, D., Parida, V., Palmié, M., & Wincent, J. (2021). How AI capabilities enable business model innovation: Scaling AI through co-evolutionary processes and feedback loops. *Journal of Business Research, 134*, 574–587.

Snihur, Y., Zott, C., & Amit, R. (2021). Managing the value appropriation dilemma in business model innovation. *Strategy Science, 6*(1), 22–38.

Soni, N., Sharma, E. K., Singh, N., & Kapoor, A. (2019). *Impact of artificial intelligence on businesses: from research, innovation, market deployment to future shifts in business models*. arXiv:1905.02092.

Tarafdar, M., Beath, C. M., & Ross, J. W. (2019). Using AI to enhance business operations. *MIT Sloan Management Review, 60*(4), 37–44.

Tecuci, G. (2012). Artificial inteligence. *WIREs Computational Statistics, 4*(2), 168–180.

Tiwana, A. (2015). Evolutionary competition in platform ecosystems. *Information Systems Research, 26*(2), 266–281.

Troilo, F. (2023). The future of human resources role: A study with business and human resources leaders in positions of regional scope in South Latin America. *Esic Market Economics and Business Journal, 54*(1), e295–e295.

Von Krogh, G. (2018). Artificial intelligence in organizations: New opportunities for phenomenon-based theorizing. *Academy of Management Discoveries, 4*(4), 404–409.

West, J. (2003). How open is open enough?: Melding proprietary and open source platform strategies. *Research policy, 32*(7), 1259–1285.

Zott, C., & Amit, R. (2017). Business model innovation: How to create value in a digital world. *NIM Marketing Intelligence Review, 9*(1), 18–23.

Artificial Intelligence and Circular Economy: What Is New for Business Model Innovation?

Óscar Montes-Pineda and Rubén Garrido-Yserte

Abstract Economic literature has described business models as the rationale for how an organization creates, delivers, and captures value; however, today, much of the research is based on linear "take-make-dispose" models rather than circular models oriented toward a more sustainable and regenerative economic system.

When circular economy principles are integrated or translated into circular business model strategies, creating and capturing value steers society's economy and resource management towards slower, more closed material and energy use cycles, leading to higher productivity. Moving from linearity to circularity involves innovation, and artificial intelligence (AI) has emerged as a significant transformative force. Empirical research suggests that companies that place sustainability issues high on the corporate agenda or strategy are also investing in the development of AI capabilities to enable circular business models (CBM), focusing on value creation through the application of solutions that reduce, reuse, and recycle material and energy resources, extend the use phase or intensify it, and even replace products with services and software solutions. Examples of how AI could help create circular business models include dynamic pricing (reducing the price of food as it approaches its expiry date to reduce food waste), matching algorithms (applicable to sharing or second-hand platforms to efficiently connect people with the things they want, from clothing to flats), or on predictive demands (applicable to reverse logistics to glass or medicine recycling). This contribution analyses how AI facilitates circular business model innovation following a prospective approach, which implies that rather than focusing on presenting a detailed review of the existing literature, it focuses more on linking relevant concepts as a way to gather valuable insights on how firms can leverage AI to advance circularity.

Ó. Montes-Pineda (✉) · R. Garrido-Yserte
Instituto Universitario de Análisis Económico y Social (IAES) y Departamento de Economía y Dirección de Empresas, Facultad de Ciencias Económicas, Empresariales y Turismo, University of Alcalá, Alcalá de Henares, Spain
e-mail: oscar.montes@uah.es; ruben.garrido@uah.es

M. T. Del Val Núñez et al. (eds.), *Artificial Intelligence and Business Transformation*, Contributions to Management Science,
https://doi.org/10.1007/978-3-031-58704-7_3

1 Introduction

The current economic model has come to be recognized as incompatible with the climate challenge, making sustainability a strategic priority for companies (Ritala et al., 2018; Ellen MacArthur Foundation, 2012, 2019). As the imperative of a greater responsibility grows, organizations can make a positive contribution through their strategy and business proposition, leading to significant consequences for current business models. Within the sustainability paradigm, the circular economy emerges as a fundamental concept, making organizations more sustainable but also competitive (Sjödin et al., 2023; Iansiti & Lakhani, 2020; Chauhan et al., 2022; Kristoffersen et al., 2020; Nishant et al., 2020).

At its core, the circular economy aims to transform the paradigm of the linear (take-make-dispose) economy towards a regenerative economy, capable of decoupling the production-consumption dynamic of finite resources from economic growth, keeping products and their materials in use for more extended periods and reducing, if not eliminating, waste and pollution, through processes of reuse, recycling or resource recovery (da Silva & Sehnem, 2022; Lacy & Rutqvist, 2015; Ellen MacArthur Foundation, 2012). As an alternative strategy, this approach comprehensively addresses today's environmental and economic challenges.

Several studies link digitalization to the successful adoption of circular industrial-business practices (Bressanelli et al., 2022; Liu et al., 2022; Neligan et al., 2022; Parida et al., 2019). Furthermore, the enabling role provided by Artificial Intelligence (AI), capable of influencing business model innovation targeting circularity, has been highlighted (Ellen MacArthur Foundation, 2019; Kristoffersen et al., 2020; Chauhan et al., 2022; Dwivedi et al., 2022; Pan & Nishant, 2023; Sjödin et al., 2023), thanks to the possibility of identifying patterns and trends, forecasting future supply and demand, and automating processes.

The potential of AI refers to its descriptive, diagnostic, predictive, and prescriptive capabilities (Madanaguli et al., 2024), which open up multiple opportunities for circularity; in the development and adoption of circular business practices and operations, the design of circular products, components, and materials, including also the optimization of infrastructures to support circular resource flows (Ellen MacArthur Foundation, 2019).

Despite its potential, the success of an AI-driven transformation toward the circular economy is far from guaranteed, and many companies fail to fully capitalize on the potential value (Minde & Bäcklund, 2023; Sjödin et al., 2023). Despite AI's enabling and accelerating role in the transition to a circular economy, investment in smart infrastructure alone is insufficient to fully realize its potential value (Sjödin & Vinit, 2021).

Successful adoption of the circular economy at the organizational level requires companies to innovate their business models, facilitating the necessary systemic change in business logic (Suchek et al., 2021). Starting a circular economy system requires adapting or creating new circular business models (hereafter, CBM), which seek to use as few resources as possible for as long as possible while maximizing value extraction (Geissdoerfer et al., 2020).

Understanding business models as a mediating construct between technology and economic value, harnessing AI capabilities would ease the redesign of business models (Iansiti & Lakhani, 2020), reframing the way how value is created, delivered, and captured to improve business circularity, meet customer needs and quality expectations, thus putting the focus on facilitating (circular) business model innovation (henceforth, CBMI). Addressing customer needs, offering solutions and services throughout the product lifecycle, removing products at the end-of-life phase, and using data and technologies from the value chain to offer more efficient solutions to the market are trends that clearly drive innovation. New business models will require optimizing the use of natural resources and reducing environmental impact through product or service innovations and other approaches toward a more circular economy.

Some real-world examples documented by the literature (Ellen MacArthur Foundation, 2019) on applications of AI to create-innovate circular business models refer to dynamic pricing (reducing the price of food as it approaches its expiry date to reduce food waste), matching algorithms (applicable to sharing or second-hand platforms to efficiently connect people with the things they want, from clothes to flats), or predictive demand estimation (applicable to reverse logistics to recycle glass and even medicines).

The above raises the question of better understanding the role of AI in circular business model innovations (CBMI) applied to circular strategies. While some of the recent studies show that digitalization is a crucial factor for the success of CBMI (Geissdoerfer et al., 2022; Neligan et al., 2022; Parida et al., 2019), they also recognize that it faces several challenges in collaboration, skills, and data (Sjödin et al., 2023; Åström et al., 2022; Antikainen et al., 2018). Other studies also recognize that it presents additional obstacles, such as a lack of trust in AI, misunderstandings about AI capabilities, and complexity that hinders interpretability and transparency (Madanaguli et al., 2024; Reim et al., 2020). CBMI is also a complex challenge in itself, with companies facing obstacles such as resistance to change, difficulty in gaining management buy-in, and lack of resources, knowledge, or skills (Geissdoerfer et al., 2022; Guldmann & Huulgaard, 2020; Tura et al., 2019).

How AI addresses sustainability issues and the objectives of the circular economy have only recently begun to receive research attention (Chauhan et al., 2022; Sjödin et al., 2023). This gap has been recognized as a research gap in a recent literature review (Madanaguli et al., 2024). Due to the relative lack of literature addressing the relationship between AI and CBMI, this chapter adopts a prospective approach to the review, which implies that rather than focusing on presenting a detailed review of the existing literature, it focuses more on linking relevant concepts as a way to gather valuable insights on how firms can leverage AI to advance circularity.

The rest of this chapter is structured as follows. The next (second) section focuses on circular business models (CBM). It starts by introducing the concept and principles of the circular economy. Then, it delves into the rationale of how companies integrate these principles into their business practices, and finishes presenting the essence of circular business models and a proposed classification. The third section examines the relationship between innovation and AI-mediated circular business models. The section presents the value of digitalization and AI in the circular economy/sustainability. It provides clarity on the concept (and processes) of AI, its current capabilities, and the future potential of AI. The section closes by exemplifying AI's impact on circular business model innovation. Section four illustrates how AI influences value creation, delivery, and capture. It also brings attention to circular business model innovation strategies, highlighting the role of AI in it. Some examples are given. The last section (five) contains the main conclusions in the form of final reflections and research trends.

2 Circular Economy and Business Models

The circular economy (CE) represents a paradigm shift in how we use natural resources and thus relate to the environment. In the new circular model, the life cycle of products, components, and materials is maintained as long as possible; waste is minimized; and resources are repeatedly reintroduced into the production cycle, creating value when goods reach the end-of-life phase (Yang et al., 2023; Ellen MacArthur Foundation, 2019). It involves moving from extracting, producing, using, and throwing away to reducing, reusing, refurbishing, recycling, and recovering.

The Ellen MacArthur Foundation (2012, 2019) recognizes that the circular economy is based on three fundamental principles. First, preserve and enhance natural capital by controlling finite stocks and balancing renewable resource flows. Second, optimize resource use by reusing products, components, and materials with maximum utility at all times, in both technical and biological cycles, and third, foster system efficiency by revealing and eliminating negative externalities. An effective circular economy strategy must optimize resources throughout the production process, from design and manufacturing through consumption and reuse, remanufacturing and recycling processes, to the disposal of non-usable waste, thus reducing resource input and waste generation.

Therefore, the circular economy seems to focus on retaining the value of products, components, and raw materials (both processed and unprocessed) in loops. This way of organizing value aims to reduce negative impacts and increase use efficiency throughout the life cycle. Business models offer several (strategic) approaches to shaping value retention that lead to extensive opportunities for value creation. Literature has studied how companies have integrated CE principles into their business practices, achieving end-customer satisfaction and describing how to capture the economic return from defined activities (Lu & Serafeim, 2023). Keeping products and materials in circulation means ensuring that products never become waste, are produced efficiently, and allow nature to regenerate itself. Circular models break with the linear approach, integrating sustainable and circular practices using, among others, recycling, reuse, and waste reduction approaches (Ellen MacArthur Foundation, 2012).

There are several ways of approaching circularity, which have come to be recognized as different models. Circular Supply Chains refer to introducing materials and energy from renewable resources. An example of this approach is General Motors,[1] which has been incorporating recycled and sustainable materials into its vehicle manufacturing processes. Product as a Service (PaaS) involves moving from a one-off relationship based on selling a tangible good to a relationship based on service. In this case, we can recall Signify's[2] Lighting-as-a-Service, formerly Philips Lightning. Extending the useful life of products means designing repairable, upgradeable, reusable, and recyclable products in all their components. The example here is Dyson,[3] which emphasizes durability and repairability in its products (hoovers, hair dryers, or air purifiers, among other products). Recycling and reuse (resource recovery) involves recovering and recycling everything that was previously wasted, which is recovered for something else. An example is IKEA.,[4] whose practices involve systematically collecting, processing, and reusing product materials at the end of their life cycle. Finally, it is worth highlighting the notion of a collaborative platform, where consumption is done collaboratively, facilitating the exchange and reuse of goods and services. The typical examples refer to companies such as Airbnb, Uber, and WeWork.[5, 6, 7]

It should be stressed that these forms or models should not be understood in isolation but rather are open to be combined, among other things, depending on the level, sector and the strategy pursued. Examples of the latter, include clean production, referring to a preventive strategy to optimize processes or eco-design as a product innovation strategy; further, industrial ecology refers to a practice of the analysis and optimization of material and energy flows or industrial symbiosis, which promotes the use of waste from one company as raw material for another.

More in detail, regarding industrial symbiosis as an example, as mentioned, it is based on a collective approach between industries pursuing a competitive advantage involving the physical exchange of materials, energy, water, and/or by-products (Chertow, 2000). The keys to industrial symbiosis are collaboration and the synergistic possibilities offered by geographical proximity. In other words, it is an industrial ecosystem in which circularity principles, such as waste restoration, reduction, and prevention, are the main value-creating activities (Geissdoerfer et al., 2020; Kristoffersen et al., 2020; Neves et al., 2020).

---

[1] GM Sustainability https://www.gm.com/commitments/sustainability

[2] Signify Circular Economy https://www.signify.com/global/sustainability

[3] Dyson, Our Journy towards Sustainability https://www.dyson.com/sustainability

[4] IKEA, Our Circular Agenda https://www.ikea.com/global/en/our-business/people-planet/

[5] WeWork Sustainability https://www.wework.com/info/impact

[6] Airbnb Sustainability https://www.airbnb.es/d/sustainabilityhosts_esp

[7] Uber Sustainability https://www.uber.com/us/en/about/sustainability/

Different industrial symbioses require different levels of coordination between actors, which affects how value is created and captured from waste (Fraccascia et al., 2019). For example, an industrial symbiosis within a company does not require coordination, as in the case of the Guitang Group, which produces sugar and diversifies its products by using its waste to create pulp, alcohol, cement, alkali, and fertilizers (Zhu et al., 2007). In contrast, an industrial symbiosis with several actors needs a high level of coordination between the partners (Fraccascia et al., 2019). An example is the industrial symbiosis in the Kalundborg region, where five companies have developed a network of symbiotic interactions involving solid waste, water, and energy exchanges (Jacobsen, 2006). Industrial symbioses can vary depending on whether a central actor is required to establish, manage, or govern the collaboration; some industrial symbioses arise naturally, and others require a central actor (Fraccascia et al., 2019).

As for the business models, literature (Fraccascia et al., 2016) has come to suggest that an industrial symbiosis consists of two main business models: one for producing waste and one for using it. Waste-producing firms can adopt either an internal exchange model using the by-product within the organization or an external exchange model in which other firms use the waste. For companies using waste, there are three possible recycling models: input substitution: less virgin input is used to obtain the same output; co-product generation: the current activity and product portfolio are improved by adding at least one new product; and new product generation: an entirely new product and activity is created. From the above, an industrial symbiosis of external exchange can lead to a set of CBM focused on associative collaboration (Minde & Bäcklund, 2023). As indicated above, this trend can be followed by different companies in different sectors.

Geissdoerfer et al. (2020) define CBM as "business models that recycle, extend, intensify and/or dematerialize material and energy loops to reduce resource inputs and waste and emissions leakage from an organizational system".

This definition underlines four generic strategies (Table 1) for creating CBM: 1) Cyclicity, referring to strategies of reuse, remanufacturing, refurbishment, reconditioning, and recycling of materials and energy within a system; 2) Extension of resource loops: prolonging the usage time of products through durable and timeless design, marketing that encourages extended use, maintenance and repair; 3) Intensification of resource cycles: maximizing the utilization of a product through solutions such as the collaborative economy; and 4) Dematerialisation: providing the functionalities of a product without hardware by replacing it with services and software solutions.

3 AI-enabled Circular Business Models

Digitalization opens new opportunities for companies to adopt CBM by offering new ways to implement sustainable yet profitable practices. Some authors (Parida et al., 2019) suggest that using digital technologies to innovate a business model

Table 1 Generic Strategies for Circular Business Models

Strategies for CBM	Definition	Examples
Cycling	Materials and energy are recycled within the system, through reuse, remanufacturing, refurbishing, and recycling.	Decommissioned industrial robots can be reused in small and medium enterprises (SMEs), where their reduced purchasing cost enables a viable way to increase productivity through automatization (Bi et al., 2015). Another example refers to IKEA, which moves towards the use of only renewable or recycled materials by 2030 (IKEA, Circular Agenda, 2024a, b).
Extending (resource loop)	Implies that the use phase of the product is extended, through long-lasting and timeless design, marketing that encourages long use phases, maintenance, and repair.	Patek Philip builds upmarket mechanical watches that last for a long time and have a timeless design that has not changed considerably over the past decades. A marketing campaign supports this with the slogan "you never actually own [this watch]. You merely look after it for the next generation" (Naas, 2015). Another example is Dyson Company, which designs and builds durable product, such as vacuum cleaners or hair dryers. (Dyson, 2024).
Intensifying (resource loop)	Implies that the use phase of the product is intensified through solutions such as sharing economy (Hamari et al., 2016) or public transport (van de Velde, 1999).	Car sharing can reduce idle times of cars and driven mileage per user significantly compared to a conventional ownership-based system (Prettenthaler and Steininger, 1999). Another example in the automotive industry is the car-rental-sharing platform Voltio in Spain, where, besides intensifying car-sharing, all vehicles are electric (See Voltio website, 2024).
Dematerializing (resource loop)	Describes the provision of product utility without hardware through substitution with service and software solutions.	These types of models refer to offering services or product-service systems instead of physical products to fulfil the same function for the user can reduce the number of produced products while enhancing the customer experience at the same time (Tan et al., 2010). An example is Spotify (See Spotify, 2024), whose offering consist in music files which can be downloaded through the internet. This offering has substituted (dematerialize) physical records (vinyl) and tapes by software files.

Source: Own elaboration based on Geissdoerfer et al. (2020)

provides new revenue streams and value-producing opportunities in industrial ecosystems (Parida et al., 2019). Digital technology fundamentally changes how industrial companies can improve productivity, increase growth, add customer value (Björkdahl, 2020), and shape business models to create, deliver, and capture value (Iansiti & Lakhani, 2014). As a result, they help the implementation of innovative and disruptive CBMs that were previously considered only conceptual (Neligan et al., 2022; Neri et al., 2023; Rosa et al., 2020).

Exemplifying the above in industrial symbiosis, literature has documented that these technologies enable optimized resource and knowledge sharing between partners, practices that are essential to improve recycling and implement more complex symbiotic systems (Colla et al., 2020; Kristoffersen et al., 2020; Liu et al., 2022). Specifically, connected sensors can support human decision-making by enabling real-time tracking, monitoring, and controlling resource flows, leading to optimized material utilization (Ingemarsdotter et al., 2019; Kristoffersen et al., 2020; Rosa et al., 2020). In addition, it can facilitate the exchange of information across industries (e.g., through digital platforms), which could help companies identify new ways to exchange resources across sectors and match supply and demand (Liu et al., 2022). Moreover, more significant human-machine interaction also improves workplace conditions. From the above, it can be argued that this group of technologies improves the economic, social, and environmental aspects of industrial symbioses (Scafà et al., 2020; Sjödin et al., 2018).

AI is often considered the pinnacle of digitalization in the collection of digital technologies due to its enormous potential to create added value (Brock & von Wangenheim, 2019). AI further amplifies the disruptive potential of digitalization by enabling systems to use sophisticated analytics, learn, generalize beyond known data, and perform tasks independently (Iansiti & Lakhani, 2020). AI can be considered an umbrella term for technologies and methods that make systems capable of performing human-like cognitive functions, including learning and reasoning (Ellen MacArthur Foundation, 2019). Moreover, pattern recognition, process optimization, prediction, and recommendation generation facilitate problem-solving, all from video, image, audio, numerical, text, and other data.

This argument has gained momentum. Madanaguli et al. (2024) refer to this potential in terms of capabilities, suggesting that AI capabilities present an extension of the capabilities provided by existing digitization capabilities. This extension is based on its ability to perform descriptive, diagnostic, predictive, and prescriptive analytics, aiding human decision-making (such as expert systems), augmenting human action (providing analytical insights based on analysis of large amounts of data), and enabling systems that can take autonomous actions (such as autonomous driving) (Kahn et al., 2020; Thomson et al., 2022). AI can generally be interpreted as the ability of a system to interpret external data correctly, learn from that data, and use those learnings (Kaplan & Haenlein, 2019; Ellen MacArthur Foundation, 2019).

Based in previous research (Geissdoerfer et al., 2020) and the work of Sjödin et al. (2023) and Sjödin and Vinit (2021), it is possible to identify different ways (see Fig. 1) in which AI can facilitate and improve CBM strategies:

- Cycling: AI can improve cycling by forecasting future material flows and improving product end-of-life strategies (Kristoffersen et al., 2020). For example, vision-based AI (sensing) can improve waste management by distinguishing between materials, enabling automatic waste audits, material sorting, and mixed

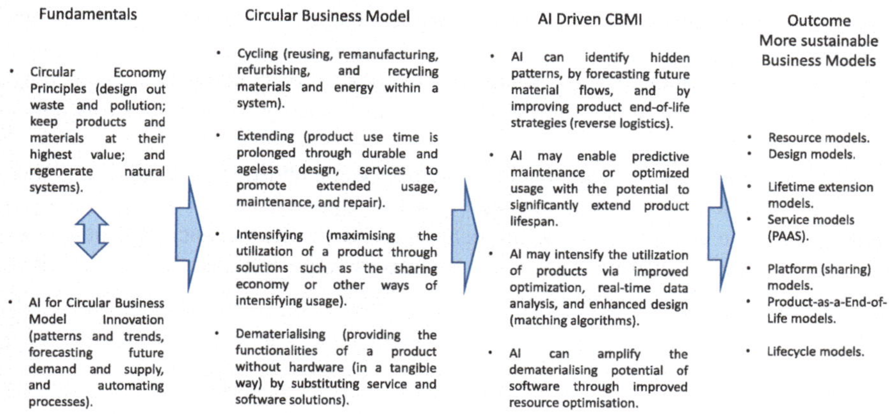

Fig. 1 AI-enabled circular business models. Source: Own elaboration based on Sjödin and Vinit (2021)

waste sorting, increasing the amount of recycled material and processing efficiency (Martinez et al., 2022; Nuñez Alonso et al., 2021).

- Extension: AI-enabled predictive and condition-based maintenance can significantly extend the lifetime of products (Liu et al., 2022). For example, Rolls-Royce offers an AI-based maintenance service that leverages data from hundreds of sensors on each aircraft engine to predict future maintenance needs, minimize delays caused by gas turbine defects, and extend utilization time between repairs (Lee et al., 2019).
- Intensification: AI can benefit the collaborative economy by improving trust, matching resources and prices, and understanding participants' preferences and attitudes (Chen et al., 2022). For example, Volvo Construction Equipment uses AI to mitigate risks and increase utilization in its rental, leasing, or performance-based contracts for heavy equipment through improved pricing and demand prediction, predictive maintenance, and intelligent inventory management (Sjödin & Vinit, 2021).
- Dematerialisation: Leveraging AI is often vital to amplify the dematerialization potential of software through better optimization of resources (Neri et al., 2023). For example, General Electric uses sensors and AI to improve wind turbines' maintenance, performance, and utilization, allowing utilities to increase their energy production capacity without adding more hardware (Iansiti & Lakhani, 2014). Another example concerns remote maintenance in the Cash Processing Systems Industry. Using sensors and AI enables online (vs staff displacement) maintenance, performance, and utilization of cash processing machinery (Giesecke+Devrient, 2024).

New evidence is growing (Kristoffersen et al., 2020; Madanaguli et al., 2024), suggesting that the increased ability to analyse and exploit data can enable and amplify

CBM. Moving from mere data collection to data analytics with AI generating better knowledge and wisdom regarding resource utilization, leading to increased resource efficiency and productivity potential. Improved knowledge allows for diagnosing how and why something has happened to a resource and discovering more efficient utilization methods.

Further, improved wisdom enables predictive capabilities to optimize future resource use and prescriptive measures that unlock dynamic and automated resource management. In other words, the application of AI further enhances the ability to make sense of data, enabling new ways to harness value and the adoption of CBM (Chauhan et al., 2022; Kristoffersen et al., 2020; Liu et al., 2022; Madanaguli et al., 2024).

4 Models Towards AI-driven Circular Business Model Innovation

There is a need to understand how the application of AI for circularity drives the transformation and innovation of companies' business models. A business model describes the design or architecture of the value creation, delivery and capture mechanisms that a company employs (Teece, 2010), usually visualized as a triangle (see Fig. 2). In other words, it describes a firm's system of activities and how they are linked to meet perceived market needs (Amit & Zott, 2012). The concept differs from strategy in that it represents how a company currently does business with customers and partners, whereas strategy describes how it intends to operate in the future (da Silva & Trkman, 2014). Therefore, continuous innovation of a company's business models is essential to execute business strategy and thus implement or enhance circular strategies through AI (Geissdoerfer et al., 2020; Zott et al., 2011).

Business model innovation consists of changing the components or their interconnections in an existing business model or creating new business models that use novel methods to create, deliver, and capture value, e.g., for a new venture or business unit of a company (Osterwalder & Pigneur, 2010; Teece, 2010). Amit and Zott (2012) recognize that business model innovation includes radical and incremental innovation, as even a slight adjustment of a business model can improve a firm's performance. The only difference between CBMI and business model innovation is that the main objective is conceptualizing and implementing CBM (Geissdoerfer et al., 2020). Therefore, CBMI can be interpreted as involving the creation of an entirely new CBM or the improvement of the circularity of existing ones through radical or minor adjustments to the methods of value creation, delivery, and capture.

Although the topic of AI and CBMI is novel, a group of authors (Minde & Bäcklund, 2023) have compiled several insights from research on business model innovation, CBMI, and AI regarding value creation, delivery, and capture.

- Value creation refers to how a firm creates its value proposition, which is the benefit a firm's product or service promises to provide to the customer (Teece,

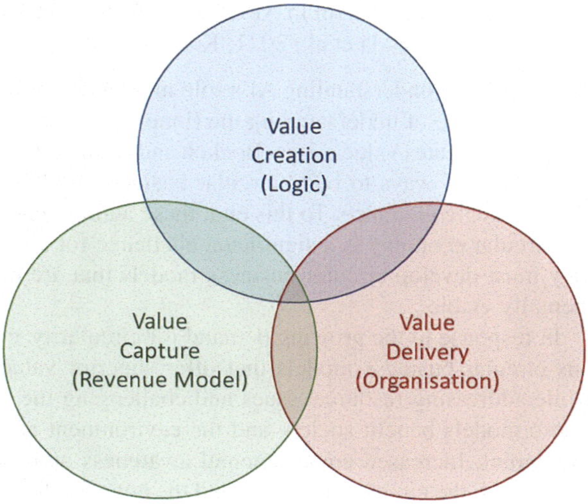

Fig. 2 AI-Business Model Triangle Visualisation

2010). A firm creates its value proposition through partnerships, key activities, and resources (Osterwalder & Pigneur, 2010). Since AI relies on large data sets, ensuring high-quality data is essential to reap the benefits of AI (Ellen MacArthur Foundation, 2019; Reim et al., 2020). As a result, data management (i.e., data collection, storage, integration, and security) and AI management become central activities in an AI-enabled CBM to acquire and use the data resource (Brock & von Wangenheim, 2019; Kristoffersen et al., 2020). These activities also require knowledgeable people, who may be internal staff or partners (Iansiti & Lakhani, 2020). Thus, partnerships, key activities, and required resources are potentially affected by AI-driven CBMI in firms' and industry ecosystems (Kanda et al., 2021; Kolagar et al., 2022; Linde et al., 2021).

- Value delivery involves establishing operational processes and activities to deliver the promised value (Teece, 2010). An enabling use case for AI is decision support to optimize the circularity of business operations (Kristoffersen et al., 2020). While ensuring that an AI solution delivers value to users is critical, enterprises often overlook the value-delivery aspect of digital solutions (Linde et al., 2021). Developing processes to ensure the adoption of AI applications and data delivery between partners are essential aspects of AI-driven CBMI in industrial ecosystems (Geissdoerfer et al., 2022; Kolagar et al., 2022; Reim et al., 2020).
- Value capture refers to how firms shape their cost structures and revenue models to convert value creation into profits (Teece, 2010). For example, AI can optimize waste streams between partners, which involves changing revenue or cost structures (Kristoffersen et al., 2020). However, companies often fail to assess how digital opportunities can alter cost and revenue structures and how value should be distributed among agents, which prevents them from capturing value (Åström et al., 2022; Linde et al., 2021). Therefore, ensuring and aligning value capture

for an entire ecosystem in AI-driven collaborative CBMI seems critical (Åström et al., 2022; Kanda et al., 2021; Kolagar et al., 2022).

Along with understanding AI's role in creating, delivering, and capturing value, the importance of understanding the (innovation) strategy under this new paradigm and how it creates value arises. Bocken and Ritala (2021) pick up on this concern by presenting six ways to build circular business models, responding to the need for clear strategic guidance. To this end, these authors understand that the transition to the circular economy is a significant challenge for all industries and companies, as they must develop circular business models that are both financially and environmentally viable.

In response to the growing demand for circularity, companies have created various circular business models that offer superior value propositions to customers while addressing resource issues and challenging the "take-make-dispose" model. These models benefit society and the environment and are sound from a business standpoint. Increased environmental awareness among the population, regulatory attention to the circular economy, and the potential for innovation and new revenues are key factors supporting the adoption of circular business models. However, as Bocken and Ritala (2021) rightly point out, clear strategic guidelines are needed to implement such initiatives successfully.

Its proposal builds on two fundamental strategic choices: innovation strategy and resource strategy. Regarding innovation strategy, companies must decide whether to improve circularity through internal processes, collaboration with external stakeholders, or a combination of both. These choices align with closed and open innovation concepts (Chesbrough, 2003). Regarding resource strategy (Bocken et al., 2016), firms must determine what type of circularity they seek: narrowing, slowing down, or closing resource loops. This strategy involves crucial decisions about managing resources and their use in the product life cycle.

Bocken and Ritala (2021, pp. 5) present these two strategies in a 2x3 matrix (see Table 2), which shows different strategic approaches to circular business models. This framework guides existing and start-up companies to initiate and manage their circular business models successfully.

AI has significant potential to amplify circular strategies and boost CBMI. Madanaguli et al. (2024) contribution completes the above strategic scheme (Table 2) by presenting, for each (resource) strategy, a circular business model based on AI. As before, it should be noted that the principles governing CBM are often implemented pursuing multiple goals and synergies in mind (Bocken et al., 2016), so a narrowing-oriented model is likely to impact slowing initiatives as well, ultimately leading to closing initiatives. The results reached by this group of authors are succinctly presented below:

- In the case of AI-enabled narrowing CBM, they recognize that AI should be leveraged to augment business analytics' descriptive and prescriptive nature, identifying inefficiencies in current processes and proposing human-understandable suggestions for better resource allocation and optimization. Such models include Siemens Mindsphere, C3.ai, and Cat Asset Intelligence. The leading AI tech-

Table 2 Circular Business Model Strategy framework

		Resource Strategy		
		Narrowing Loops	Slowing Loops	Closing Loops
Innovation Strategy	**Open**	**Value proposition (example):** Reduce waste and resources in design and production processes. **Value creation and delivery:** Reduce cost and negative impact through new technologies and processes in collaboration with suppliers, customers, and others. **Value capture logic:** Save cost and resources. **Case examples:** Industry collaboration on cleaner refrigeration technology; Sony and Samsung collaboration on LCD efficiency.	**Value proposition (example):** Reuse resources to broaden the offerings to the customer (e.g., vintage, second-hand). **Value creation and delivery:** Create value by connecting internal and external resource flows via generative models. **Value capture logic:** Increase the number of transactions in an ecosystem via the reuse of products. **Case examples:** H&M—Sellpy collaboration on second-hand clothes market; iFixit repair platform; ResQ Club excess food sales model.	**Value proposition (example):** A circular offering that involves lower environmental footprint and resource burden. **Value creation and delivery:** Combine resource flows from external ecosystem into customer offerings. **Value capture logic:** Lower the cost of resources used in customer offerings, improve brand and corporate image. **Case examples:** Interface Networks for 'circular carpets' with ZSL, Aquafil, and fishery communities; JLR and Novelis closing the aluminum loop.
	Closed	**Value proposition (example):** Reduce waste and resources in design and production processes. **Value creation and delivery:** Reduce cost and negative impact through internal technology, process, and design innovations. **Value capture logic:** Save cost and resources. **Case examples:** Companies like Apple minimize packaging and using recycled materials; McDonald's "fried for fuel."	**Value proposition (example):** High-quality products with high customer value. **Value creation and delivery**: Long-lasting design and repair services; Create more value from less resources. **Value capture logic:** Price premium through achieving quality leadership and customer loyalty; create value from the same product multiple times. **Case examples:** Long-life warranties; hotel linen rental services focused on product longevity.	**Value proposition (example):** Connect with customers by using, recovering, and maintaining post-consumer materials. **Value creation and delivery:** Increase customer retention and repurchases via take-back plans. **Value capture logic:** Resource efficiency, improve brand and reputation, reduce the cost of materials. **Case example:** Take-back, rental, and lease models to recover the company's materials, such as MUD Jeans Lease and Philips pay per lux.

Source: Bocken and Ritala (2021, pp. 5)

nologies that can be employed are data analytics, machine learning, simulation, decision systems, and generative AI.

- The second AI-based slowdown CBM proposes using AI to improve diagnostics and predictive data analytics to augment and automate design and delivery processes through data-driven designs that are durable and pre-emptively repaired before failures occur by analyzing usage patterns. Examples of such models include Siemens Healthineers, Bosch Maintenance 4.0, and ABB. The key AI technologies include data analytics, machine learning, digital twins, data-driven design, and predictive maintenance.
- The third AI-based closing CBM recognizes Al′s descriptive, diagnostic, predictive, and prescriptive capabilities with partners through new business model configurations that use data-driven design to enable easy assembly and disassembly while identifying the secondary market for reuse and refurbished parts through customer matching. Examples can be seen at Xerox, Dell, and Renault Refactory. Key AI technologies relate to data-driven design, digital twins, analytics and forecasting, and generative AI.

5 Conclusion

This chapter has explored AI's potential to boost Circular Business Model Innovations (CBMI) within circular business strategies. The discussion encompasses leveraging existing CBM practices with AI to unlock new value creation, delivery, and capturing dimensions. The potential of digitalization and AI in the circular economy/sustainability relies on the increased ability to analyze and exploit data, capable of enabling and amplifying CBM.

Moving from mere data collection to data analytics with AI generates better knowledge and wisdom regarding resource utilization, leading to increased resource efficiency and productivity potential. Improved knowledge allows for diagnosing how and why something has happened to a resource and discovering more efficient utilization methods. Better wisdom enables predictive capabilities to optimize future resource use and prescriptive measures that unlock dynamic and automated resource management. In other words, the application of AI further enhances the ability to make sense of data, enabling new ways to harness value and the adoption of CBM (Chauhan et al., 2022; Kristoffersen et al., 2020; Liu et al., 2022; Madanaguli et al., 2024).

Despite the recognized potential of digitalization and the promising prospects of AI in fostering the transition toward circularity, it is apparent that the domain of AI-driven CBMI remains underexplored and demands further research (Madanaguli et al., 2024; Bressanelli et al., 2022; Chauhan et al., 2022; Kristoffersen et al., 2020; Liu et al., 2022; Neligan et al., 2022; Parida et al., 2019).

Findings offer theoretical insights into implementing AI-driven CBMI through concrete strategies such as narrowing, slowing, and closing loops. While evidence suggests that AI can enhance circularity and articulate principles to drive it, there is a need to identify new routines and processes guided by AI to establish a comprehensive framework.

Throughout various sections, we acknowledge the significance of Circular Economy (CE) guiding principles in shaping corporate strategy (and in turn, business models) and facilitating value creation through AI application (Iansiti & Lakhani, 2020; Reim et al., 2020; Sjödin et al., 2021). In line with literature (Kolagar et al., 2022), innovation, collaboration, and business model alignment are essential for the digital transformation of industrial/business ecosystems. However, a deeper understanding of the application of AI in diverse industrial/business ecosystems is essential.

Critical questions persist as we delve into the various facets of AI in CBMI. The actual business value of AI for Circular Business Models (CBM) and its impact on firm performance remain unclear. The latter prompts consideration of whether CBM acts as a mediator or moderator in the relationship between AI adoption and the firm—additionally, the mechanisms governing the interaction of AI within firm environments for CBM lack of clarity. The strategies and activities involved in enabling AI for CBM innovation need to be better understood. These gaps underscore the necessity for further exploration and comprehension to enhance our understanding of the intricate dynamics between AI and CBM innovation in business settings.

References

Amit, R., & Zott, C. (2012). Creating value through business model innovation. *MIT Sloan Management Review, 53*(3), 45–49. https://search.proquest.com/docview/963962187

Antikainen, M., Uusitalo, T., & Kivikytö-Reponen, P. (2018). Digitalisation as an enabler of circular economy. *Procedia CIRP, 73*, 45–49. https://doi.org/10.1016/j.procir.2018.04.027

Åström, J., Reim, W., & Parida, V. (2022). Value creation and value capture for AI business model innovation: A three-phase process framework. *Review of Managerial Science, 16*(7), 2111–2133. https://doi.org/10.1007/s11846-022-00521-z

Bi, Z. M., Liu, Y., Baumgartner, B., Culver, E., Sorokin, J. N., Peters, A., Cox, B., Hunnicutt, J., Yurek, J., & O'Shaughnessey, S. (2015). Reusing industrial robots to achieve sustainability in small and medium-sized enterprises (SMEs). *Industrial Robot, 42*(3), 264–273. https://doi.org/10.1108/IR-12-2014-0441

Björkdahl, J. (2020). Strategies for digitalization in manufacturing firms. *California Management Review, 62*(4), 17–36. https://doi.org/10.1177/0008125620920349

Bocken, N., & Ritala, P. (2021). Six ways to build circular business models. *Journal of Business Strategy*, Vol. ahead-of-print No. ahead-of-print. https://doi.org/10.1108/JBS-11-2020-0258

Bocken, N. M. P., de Pauw, I., Bakker, C., & van der Grinten, B. (2016). Product design and business model strategies for a circular economy. *Journal of Industrial and Production Engineering, 33*(5), 308–320. https://doi.org/10.1080/21681015.2016.1172124

Bressanelli, G., Adrodegari, F., Pigosso, D. C. A., & Parida, V. (2022). Towards the smart circular economy paradigm: A definition, conceptualization, and research agenda. *Sustainability (Basel, Switzerland), 14*(9), 4960. https://doi.org/10.3390/su14094960

Brock, J. K., & von Wangenheim, F. (2019). Demystifying AI: What digital transformation leaders can teach you about realistic artificial intelligence. *California Management Review, 61*(4), 110–134. https://doi.org/10.1177/1536504219865226

Chauhan, C., Parida, V., & Dhir, A. (2022). Linking circular economy and digitalisation technologies: A systematic literature review of past achievements and future promises. *Technological Forecasting and Social Change, 177*, 121508. https://doi.org/10.1016/j.techfore.2022.121508

Chen, Y., Prentice, C., Weaven, S., & Hsiao, A. (2022). A systematic literature review of AI in the sharing economy. *Journal of Global Scholars of Marketing Science, 32*(3), 434–451. https://doi.org/10.1080/21639159.2020.1808850

Chertow, M. R. (2000). Industrial symbiosis: Literature and taxonomy. *Annual Review of Energy and the Environment, 25*(1), 313–337. https://doi.org/10.1146/annurev.energy.25.1.313

Chesbrough, H. W. (2003). *Open innovation: The new imperative for creating and profiting from technology*. Harvard Business Press.

Colla, V., Pietrosanti, C., Malfa, E., & Peters, K. (2020). Environment 4.0: How digitalization and machine learning can improve the environmental footprint of the steel production processes. *Matériaux & Techniques, 108*(4), 507. https://doi.org/10.1051/mattech/2021007

da Silva, C. M., & Trkman, P. (2014). Business model: What it is and what it is not. *Long Range Planning, 47*(6), 379–389. https://doi.org/10.1016/j.lrp.2013.08.004

da Silva, T. H. H., & Sehnem, S. (2022). The circular economy and industry 4.0: Synergies and challenges. *Revista de Gestão, 29*, 300–313.

Dyson, J. (2024). *Our journey towards sustainability*. Video link https://www.dyson.com/sustainability

Dwivedi, A., Moktadir, M. A., Jabbour, C. J. C., & de Carvalho, D. E. (2022). Integrating the circular economy and industry 4.0 for sustainable development: Implications for responsible footwear production in a big data-driven world. *Technological Forecasting and Social Change, 175*, 121335.

Ellen MacArthur Foundation. (2012). *Towards a circular economy-economic and business rationale for an accelerated transition*. https://www.ellenmacarthurfoundation.org/towards-a-circular-economy-business-rationale-for-an-accelerated-transition

Ellen MacArthur Foundation. (2019). *Artificial intelligence and the circular economy - AI as a tool to accelerate the transition*. https://www.ellenmacarthurfoundation.org/artificial-intelligence-and-the-circular-economy

Fraccascia, L., Giannoccaro, I., & Albino, V. (2019). Business models for industrial symbiosis: A taxonomy focused on the form of governance. *Resources, Conservation and Recycling, 146*, 114–126. https://doi.org/10.1016/j.resconrec.2019.03.016

Fraccascia, L., Magno, M., & Albino, V. (2016). Business models for industrial symbiosis: A guide for firms. *Procedia Environmental Science, Engineering and Management, 3*(2), 83–93.

Geissdoerfer, M., Pieroni, M. P. P., Pigosso, D. C. A., & Soufani, K. (2020). Circular business models: A review. *Journal of Cleaner Production, 277*, 123741. https://doi.org/10.1016/j.jclepro.2020.123741

Geissdoerfer, M., Santa-Maria, T., Kirchherr, J., & Pelzeter, C. (2022). Drivers and barriers for circular business model innovation. *Business Strategy and the Environment*. https://doi.org/10.1002/bse.3339

Giesecke+Devrient. (2024). Sustainability Report 2023 https://www.gi-de.com/en/group/sustainability

Guldmann, E., & Huulgaard, R. D. (2020). Barriers to circular business model innovation: A multiple-case study. *Journal of Cleaner Production, 243*, 118160. https://doi.org/10.1016/j.jclepro.2019.118160

Hamari, J., Sjöklint, M., & Ukkonen, A. (2016). The sharing economy: Why people participate in collaborative consumption. *Journal of the Association for Information Science and Technology, 67*(9), 2047–2059. https://doi.org/10.1002/asi.23552

Iansiti, M., & Lakhani, K. R. (2014). Digital ubiquity: How connections, sensors, and data are revolutionizing business. *Harvard Business Review*, 92.

Iansiti, M., & Lakhani, K. R. (2020). Competing in the age of AI: How machine intelligence changes the rules of business. *Harvard Business Review*, 98.

IKEA. (2024a). Circular Agenda. Webpage link https://www.ikea.com/global/en/our-business/people-planet/our-circularagenda/

IKEA. (2024b). *Why the future of furniture is circular*. Video link https://youtu.be/EvfNA0FtesY?si=LuzUiRljLpMTEdiK

Ingemarsdotter, E. K., Jamsin, E., Kortuem, G. W., & Balkenende, A. R. (2019). Circular strategies enabled by the internet of things: A framework and analysis of current practice. *Sustainability, 11*(20), 5689. https://doi.org/10.3390/su11205689

Jacobsen, N. B. (2006). Industrial Symbiosis in Kalundborg, Denmark: A quantitative assessment of economic and environmental aspects. *Journal of Industrial Ecology, 10*(1–2), 239–255. https://doi.org/10.1162/108819806775545411

Kahn, L. H., Savas, O., Morrison, A., Shaffer, K.A., & Zapata, L. (2020). Modelling hybrid human-artificial intelligence cooperation: a call center customer service case study. In *Proceedings - 2020 IEEE International Conference on Big Data, Big Data 2020*, pp. 3072–3075. https://doi.org/10.1109/BIGDATA50022.2020.9377747.

Kanda, W., Geissdoerfer, M., & Hjelm, O. (2021). From circular business models to circular business ecosystems. *Business Strategy and the Environment, 30*(6), 2814–2829. https://doi.org/10.1002/bse.2895

Kaplan, A., & Haenlein, M. (2019). Siri, Siri, in my hand: Who's the fairest in the land? On the interpretations, illustrations, and implications of artificial intelligence. *Business Horizons, 62*(1), 15–25. https://doi.org/10.1016/j.bushor.2018.08.004

Kolagar, M., Parida, V., & Sjödin, D. (2022). Ecosystem transformation for digital servitization: A systematic review, integrative framework, and future research agenda. *Journal of Business Research, 146*, 176–200. https://doi.org/10.1016/j.jbusres.2022.03.067

Kristoffersen, E., Blomsma, F., Mikalef, P., & Li, J. (2020). The smart circular economy: A digital-enabled circular strategies framework for manufacturing companies. *Journal of Business Research, 120*, 241–261. https://doi.org/10.1016/j.jbusres.2020.07.044

Lacy, P., & Rutqvist, J. (2015). *Waste to wealth: The circular economy advantage*. Springer.

Lee, S. M., Lee, D., & Kim, Y. S. (2019). The quality management ecosystem for predictive maintenance in the Industry 4.0 era. *International Journal of Quality Innovation, 5*, 4. https://doi.org/10.1186/s40887-019-0029-5

Linde, L., Sjödin, D., Parida, V., & Gebauer, H. (2021). Evaluation of digital business model opportunities. *Research-Technology Management, 64*(1), 43–53. https://doi.org/10.1080/08956308.2021.1842664

Liu, Q., Trevisan, A. H., Yang, M., & Mascarenhas, J. (2022). A framework of digital technologies for the circular economy: Digital functions and mechanisms. *Business Strategy and the Environment, 31*(5), 2171–2192. https://doi.org/10.1002/bse.3015

Lu, S., & Serafeim, G. (2023). How AI will accelerate the circular economy. *Environmental Sustainability, Harvard Business Review*. https://hbr.org/2023/06/how-ai-will-accelerate-the-circular-economy

Madanaguli, A., Sjödin, D., Parida, V., & Mikalef, P. (2024). Artificial intelligence capabilities for circular business models: Research synthesis and future agenda. *Technological Forecasting and Social Change, 200*, 123189, ISSN 0040-1625. https://doi.org/10.1016/j.techfore.2023.123189

Martinez, P., Mohsen, O., Al-Hussein, M., & Ahmad, R. (2022). Vision-based automated waste audits: A use case from the window manufacturing industry. *International Journal of Advanced Manufacturing Technology, 119*(11–12), 7735–7749. https://doi.org/10.1007/s00170-022-08730-2

Minde, B. T., & Bäcklund, N. (2023). *Unlocking the potential of AI-driven circular business model innovation: A case study of an industrial symbiosis*. Luleå University of Technology.

Naas, R. (2015). *Patek Philippe celebrates 20 Years of its iconic advertising campaign*. Forbes Online Dec. Link https://www.forbes.com/sites/robertanaas/2016/12/09/patek-philippe-celebrates-20-years-of-its-iconic-advertisingcampaign-you-never-actually-own-a-patek-philippe/?sh=40b39774475b

Neligan, A., Baumgartner, R. J., Geissdoerfer, M., & Schöggl, J. (2022). Circular disruption: Digitalisation as a driver of circular economy business models. *Business Strategy and the Environment, 32*(3), 1175–1188. https://doi.org/10.1002/bse.3100

Neri, A., Negri, M., Cagno, E., Franzò, S., Kumar, V., Lampertico, T., & Bassani, C. A. (2023). The role of digital technologies in supporting the implementation of circular economy practices by industrial small and medium enterprises. *Business Strategy and the Environment*, 1–26. https://doi.org/10.1002/bse.3388

Neves, A., Godina, R., Azevedo, S. G., & Matias, J. C. O. (2020). A comprehensive review of industrial symbiosis. *Journal of Cleaner Production, 247*, 119113. https://doi.org/10.1016/j.jclepro.2019.119113

Nishant, R., Kennedy, M., & Corbett, J. (2020). Artificial intelligence for sustainability: Challenges, opportunities, and a research agenda. *International Journal of Information Management, 53*, 102104. https://doi.org/10.1016/j.ijinfomgt.2020.102104

Nuñez Alonso, S. L., Reier Forradellas, R. F., Pi Morell, O., & Jorge-Vazquez, J. (2021). Digitalization, circular economy and environmental sustainability: The application of artificial intelligence in the efficient self-management of Waste. *Sustainability (Basel, Switzerland), 13*(4), 2092. https://doi.org/10.3390/su13042092

Osterwalder, A., & Pigneur, Y. (2010). *Business model generation: A handbook for visionaries, game changers, and challengers*. John Wiley & Sons.

Pan, S. L., & Nishant, R. (2023). Artificial intelligence for digital sustainability: An insight into domain-specific research and future directions. *International Journal of Information Management, 72*, 102668.

Parida, V., Sjödin, D., & Reim, W. (2019). Reviewing literature on digitalization, business model innovation, and sustainable industry: Past achievements and future promises. *Sustainability, 11*(2), 391. https://doi.org/10.3390/su11020391

Prettenthaler, F. E., & Steininger, K. W. (1999). From ownership to service use lifestyle: The potential of car sharing. *Ecological Economics, 28*(3), 443–453. https://doi.org/10.1016/S0921-8009(98)00109-8

Reim, W., Åström, J., & Eriksson, O. (2020). Implementation of artificial intelligence (AI): A roadmap for business model innovation. *AI, 1*(2), 180–191. https://doi.org/10.3390/ai1020011

Ritala, P., Huotari, P., Bocken, N., Albareda, L., & Puumalainen, K. (2018). Sustainable business model adoption among S&P 500 firms: A longitudinal content analysis study. *Journal of Cleaner Production, 170*, 216–226. https://doi.org/10.1016/j.jclepro.2017.09.159

Rosa, P., Sassanelli, C., Urbinati, A., Chiaroni, D., & Terzi, S. (2020). Assessing relations between circular economy and Industry 4.0: A systematic literature review. *International Journal of Production Research, 58*(6), 1662–1687. https://doi.org/10.1080/00207543.2019.1680896

Scafà, M., Marconi, M., & Germani, M. (2020). A critical review of symbiosis approaches in the context of Industry 4.0. *Journal of Computational Design and Engineering, 7*(3), 269–278. https://doi.org/10.1093/jcde/qwaa022

Sjödin, D., Parida, V., Petrovic, A., & Leksell, M. (2018). Smart factory implementation and process innovation. *Research-Technology Management, 61*(5), 22–31. https://doi.org/10.1080/08956308.2018.1471277

Sjödin, D., & Vinit, P. (2021). AI-driven circular business models — Leveraging AI for circularity benefits in manufacturing. *Managment of Innovation and Technology, 18*, 10–11. https://mgmt.imit.se/artiklar/ai-driven-circular-business-models/

Sjödin, D., Vinit, P., & Kohtamäki, M. (2023). Artificial intelligence enabling circular business model innovation in digital servitization: Conceptualizing dynamic capabilities, AI capacities, business models and effects. *Technological Forecasting and Social Change, 197*, 122903. https://doi.org/10.1016/j.techfore.2023.122903

Spotify. (2024). Equity and Impact Report 2023 https://www.lifeatspotify.com/reports/Spotify-Equity-Impact-Report-2023.pdf

Suchek, N., Fernandes, C. I., Kraus, S., Filser, M., & Sjögrén, H. (2021). Innovation and the circular economy: A systematic literature review. *Business Strategy and the Environment, 30*(8), 3686–3702. https://doi.org/10.1002/bse.2834

Tan, A. R., Matzen, D., McAloone, T. C., & Evans, S. (2010). Strategies for designing and developing services for manufacturing firms. *CIRP Journal of Manufacturing Science and Technology, 3*(2), 90–97. https://doi.org/10.1016/j.cirpj.2010.01.001

Teece, D. J. (2010). Business models, business strategy and innovation. *Long Range Planning, 43*(2), 172–194. https://doi.org/10.1016/j.lrp.2009.07.003

Thomson, L., Kamalaldin, A., Sjödin, D., & Parida, V. (2022). A maturity framework for autonomous solutions in manufacturing firms: The interplay of technology, ecosystem, and business model. *The International Entrepreneurship and Management Journal, 18*(1), 125–152. https://doi.org/10.1007/s11365-020-00717-3

Tura, N., Hanski, J., Ahola, T., Ståhle, M., Piiparinen, S., & Valkokari, P. (2019). Unlocking circular business: A framework of barriers and drivers. *Journal of Cleaner Production, 212*, 90–98. https://doi.org/10.1016/j.jclepro.2018.11.202

van de Velde, D. M. (1999). Organisational forms and entrepreneurship in public transport: Classifying organisational forms. *Transport Policy, 6*(3), 147–157. https://doi.org/10.1016/S0967-070X(99)00016-5

Voltio (2024). Movilidad Sostenible https://voltio.com/movilidad-sostenible/

Yang, C., Ma, H., Liu, K. H., & Yuan, M. H. (2023). Measuring circular economy transition potential for industrial wastes. *Sustainable Production and Consumption*. https://doi.org/10.1016/j.spc.2023.06.013

Zhu, Q., Lowe, E. A., Wei, Y., & Barnes, D. (2007). Industrial symbiosis in China: A case study of the Guitang Group. *Journal of Industrial Ecology, 11*(1), 31–42. https://doi.org/10.1162/jiec.2007.929

Zott, C., Amit, R., & Massa, L. (2011). The business model: Recent developments and future research. *Journal of Management, 37*(4), 1019–1042. https://doi.org/10.1177/0149206311406265

Artificial Intelligence and Sustainability

Juan Piñeiro-Chousa, M. Ángeles López Cabarcos, Noelia Romero-Castro, and Isaac González-López

Abstract Society faces great challenges. Achieving Sustainable Development (SD) is a priority that has earned the commitment of governments following widespread adoption of the United Nations (UN) 2030 Agenda and its 17 Sustainable Development Goals (SDGs). Artificial Intelligence (AI) and the associated digital technologies are driving major changes in many areas of socioeconomic systems, presenting both opportunities and risks for sustainability. There is extensive academic literature examining the contribution of AI to SD. After a brief review of this literature, this chapter provides a complementary approach, focusing on the analysis of online news and interaction with two chatbots to question AI itself about its role in achieving sustainability. Implications are derived for researchers, policymakers and private economic actors that should acknowledge and consciously consider the potential conflicts between AI and SD.

Supplementary Information The online version contains supplementary material available at https://doi.org/10.1007/978-3-031-58704-7_4.

J. Piñeiro-Chousa (✉) · N. Romero-Castro
Department of Finance and Accounting, University of Santiago de Compostela, Santiago, Spain
e-mail: j.pineiro@usc.es; noe.romero@usc.es

M. Á. L. Cabarcos
Department of Business Administration, University of Santiago de Compostela, Santiago, Spain
e-mail: angeles.lopez.cabarcos@usc.es

I. González-López
University of Santiago de Compostela, Santiago, Spain
e-mail: isaacjose.gonzalez.lopez@usc.es

M. T. Del Val Núñez et al. (eds.), *Artificial Intelligence and Business Transformation*, Contributions to Management Science,
https://doi.org/10.1007/978-3-031-58704-7_4

1 Introduction

Digital technologies are radically transforming society, and it is unclear how society will respond to such rapid and dramatic changes. At the same time, the world has recognized the need to transform its development patterns to achieve sustainable development (SD) by adopting the 17 Sustainable Development Goals (SDGs) of the UN 2030 Agenda. It is reasonable to expect digital technologies to boost global sustainability (Fan et al., 2023; Truby, 2020). However, so far, there is little evidence that digital technologies have made a genuine contribution to sustainable development (Del Río Castro et al., 2021). Digital technologies offer both opportunities and risks (Yeh et al., 2021). Novel data sources, enhanced analytical capabilities, and collaborative digital ecosystems have brought about the rise of artificial intelligence (AI), blockchain, and the Internet of things (IoT), with valuable applications in the pursuit of sustainability (Wu, Shirkey, et al., 2022). However, they also entail controversies in terms of social and environmental sustainability (Goralski & Tan, 2020; Scholz et al., 2018). Thus, the scientific community and the public have started to demand the assessment of technological advancements in terms of efficiency, ethics, and sustainability (Vinuesa et al., 2020), acknowledging that sustainability-related issues cannot be solved by technology alone (Del Río Castro et al., 2021).

The emergence of AI and its increasing impact on many sectors means that evaluating its effect on achieving sustainability is important (Vinuesa et al., 2020). Digital technologies and AI are disruptive technologies that users combine to try to solve social and economic challenges. However, they do not offer a magic cure (Meitei et al., 2023). They should be used responsibly and thoughtfully (Goralski & Tan, 2020; Palomares et al., 2021).

A hot topic in academic research in the last few years has been the relationship between AI and sustainable development. This topic has received increasing attention in the scientific literature, and there are already many literature reviews on the matter. Yeh et al. (2021) offered a new perspective, analyzing the public perception of AI and its connections to the SDGs. This public perception is greatly influenced by social media, including online news. AI reaches the public through open AI text generation tools such as ChatGPT and Bard. The role of chatbots in higher education learning, teaching, and assessment is a hotly debated issue (Rudolph et al., 2023). Online news and chatbots can exert a direct impact on the creation of public opinion that can condition social resilience (Vinuesa et al., 2020) to the fast changes of today. Given this direct impact, it is of interest to analyze how the relationship between AI and sustainability is presented in online news and chatbots. This chapter provides such analysis, offering an innovative approach that complements the vast scientific output found in the academic literature.

2 Artificial Intelligence and Sustainability

2.1 AI

Although the concept of AI was coined by John McCarthy in 1955 (McCarthy et al., 2006), it does not have a universal definition. Vinuesa et al. (2020) offered an operative definition, identifying specific AI capabilities: perception, decision-making, prediction, automatic knowledge extraction and pattern recognition from data, interactive communication, and logical reasoning. AI needs data flows to learn and requires other digital technologies to acquire, store, compute, and share that data (Palomares et al., 2021; Pigola et al., 2021).

AI encompasses multiple applications in diverse fields, from smart cities, supply chains, energy (Wu, Shirkey, et al., 2022), to accounting (Peng et al., 2023), project management (Nieto-Rodriguez & Viana Vargas, 2023), circular economy (Lu & Serafeim, 2023), or sustainable finance (Al-Sartawi et al., 2022). It affects global productivity, competitiveness, and many other dimensions of society, such as equality and inclusion (Kar et al., 2022; Vinuesa et al., 2020). However, the vast potential of AI is repeatedly overstated. There are also possible negative effects of AI on sustainable development (Meitei et al., 2023; Sætra, 2021). Denning and Denning (2020) proposed 10 ethical and design dilemmas posed by AI: explainability, fragility, bias, fakes, high costs of reliable training data, military use of AI, weapons and control, employment and jobs, surveillance capitalism, and decision making. Bachmann et al. (2022) highlight problems related to data gaps, data biases, high energy consumption of computational resources, ethical concerns, privacy, ownership, and security issues. Among the negative environmental impacts (Ligozat et al., 2022), special mention must be made to the high carbon footprint (Rolnick et al., 2023; Wu, Raghavendra, et al., 2022), since AI training and operation demands significant energy consumption (Nieto-Rodriguez & Viana Vargas, 2023). It can also result in inequalities due to the restriction of educational and computing resources throughout the world (Pigola et al., 2021; Sætra, 2021; Vinuesa et al., 2020), it presents data ownership and privacy dilemmas (Vinuesa et al., 2020), and it poses major risks for individuals, their values, and their personal development (Fernández-Aller et al., 2021).

2.2 Sustainability and the SDGs

SD, defined in the Brundtland Report (WCED, 1987), seeks a balance across three pillars: social sustainability, economic sustainability, and environmental sustainability (Purvis et al., 2018). In 2015, the UN set 17 Sustainable Development Goals (SDGs) covering environmental, social, and economic aims. They are divided into 169 targets and 232 indicators that were internationally ratified in the 2030 Agenda for Sustainable Development (UN, 2015). Allen et al. (2018) showed that while good progress has been made in the implementation stage, there are large gaps in further stages. Halkos and Gkampoura (2021) found that although economy-related targets are close to being achieved, those related to education, cities, community sustainability, and climate change lag behind. The latest report by the United Nations (UN, 2015) offers an urgent warning about global warming and climate change.

The linkages between the SDGs are notable (Sætra, 2021). The SDGs are interconnected and should be treated as a cohesive system (Yeh et al., 2021). One key criticism of the SDGs is the potential disconnect between combating climate change and achieving the rest of the SDGs. Fuso Nerini et al. (2019) concluded that although climate change may jeopardize the achievement of 16 SDGs, taking action against climate change may undermine 12 of these goals. Del Río Castro et al. (2021) noted that the SDGs present several research gaps: flawed understanding of complexities and interlinkages; design shortcomings and imbalances; implementation and governance hurdles; unsuitable indicators and assessment methodologies; truncated adoption and off-target progress; unclear responsibilities and a lack of coordination; and the untapped role of technological innovation and knowledge management.

2.3 The Relation Between AI and SD in Academic Literature

In recent years, several studies have examined the opportunities and challenges of AI for sustainable development and the SDGs based on academic literature reviews. Despite not focusing specifically on sustainable development, Chui et al. (2018) provided one of the first comprehensive and evidence-based accounts of the beneficial use of AI based on hundreds of use cases across different industries. Some studies linking AI and sustainable development have examined the impacts of AI across the broad areas of economic, social, and environmental sustainability (Vinuesa et al., 2020). Others have adopted a more targeted approach. For example, Khakurel et al. (2018) reviewed the potential long-term impacts of AI on sustainable development, finding positive and negative effects across five levels. Palomares et al. (2021) described six categories or perspectives of human need, presenting a SWOT analysis of AI for achieving the SDGs and offering recommendations on how AI can help in this regard.

Scholars have also provided in-depth analysis of each one of the 17 SDGs (Bachmann et al., 2022; Pigola et al., 2021), with others focusing on a single SDG. For example, Di Vaio et al. (2020) mainly considered the impact of sustainable business models on SDG 12 (Responsible consumption and production). Some studies have focused on a limited geographic scope. For instance, Pigola et al. (2021) performed a public survey to rank preferable AI-driven digital technologies for different SDGs in Brazil and Portugal. Other studies have focused on the analysis of certain industries. For example, Goralski and Tan (2020) used three case studies in the water management, agriculture, and sanitation services sectors.

Vinuesa et al. (2020) offered one of the most comprehensive reviews to date on the relationship between AI and the SDGs. Applying a consensus-based expert elicitation process to identify the possibilities of synergies and trade-offs of AI, they found that AI can empower 134 of the 169 targets in the 2030 Agenda, while inhibiting 59. More recently, Kar et al. (2022) systematically reviewed the impacts of AI on sustainability and identified the range of main algorithms used, the opportunities and barriers to the contribution of AI to sustainable development, and some use cases by industry. Bachmann et al. (2022) conducted a literature review and content analysis to show that data-driven technologies enable the achievement of the 17 SDGs, although advances are also hindered by problems. Finally, Meitei et al. (2023) reviewed the application of AI and machine learning (ML) techniques in the academic literature on each of the SDGs, concluding that AI holds promise but that there is overexuberance about its positive outcome.

The academic literature shows a general inclination to present AI as an enabler of sustainability. Vinuesa et al. (2020) pointed out that this bias can be explained by two main reasons: (1) self-interest toward publishing positive results and (2) the short life of AI developments, which limits the availability of longer-term studies capable of discovering detrimental aspects of AI. The UN-driven initiative AI4Good is one example of how AI is seen as a force for good (International Telecommunication Union, 2023), establishing interdisciplinary partnerships to foster the development of AI applications to meet the SDGs (Tomašev et al., 2020). To support the development of AI that is aware of its environmental implications, Schwartz et al. (2020) proposed the concept of Green AI. They describe Red AI as AI that has resulted in escalating computational (and thus carbon) costs, whereas Green AI refers to AI that yields novel results while encouraging a reduction in resources spent. However, Ligozat et al. (2022) noted that the current Green AI literature only considers a small proportion of the direct environmental impacts, suggesting the need to use life cycle assessment in design and development and focusing on full net environmental impacts, not only those related to climate change. To address the high complexity, costs, and carbon footprint of AI applications in project management, green algorithms are being developed, such as digital twins to predict and streamline logistics costs of wind parks (Nieto-Rodriguez & Viana Vargas, 2023).

As this review shows, the academic literature provides considerable evidence of the positive impacts of AI on sustainable development. However, some concerns must also be addressed. To complement the approaches found in the academic literature, the following sections of this chapter provide analysis of online news and chatbot discourses, describing how the relationship between AI and sustainable development is presented to the public and can contribute to forming public opinion.

3 A New Approach to the Analysis of the Contribution of AI to SD

The previous section shows the abundance of articles on the intersection of AI and sustainability (Palomares et al., 2021), fueled by the rapid development of AI. It therefore seems reasonable to seek other sources to develop a broader up-to-date view of the topic. Despite its possible bias and lack of full industry representation (Frijters & Velamuri, 2010), online news can reveal trends, topics of interest, and other insights (Nghiem et al., 2016; Vermeer et al., 2020).

Google News is a news aggregator service that covers a wide range of news sources. It was chosen for this study to mitigate possible topic bias due to the use of a single source (Patankar & Bose, 2016). To analyze news content, an online service called Graphext was used (https://www.graphext.com/). This service combines several natural language processing tools such as keyword extraction, entity extraction, sentiment analysis, and topic clustering. It also offers graphical visualizations for these analyses. It has been applied for science mapping in academic literature reviews of diverse topics (Núñez-Merino et al., 2022).

In addition, to examine how AI itself reports on its capacity to contribute to sustainable development, the Bard and ChatGPT tools were questioned. These two chatbots are among the most used in academic research to compare their applications and usefulness in different areas (Gan et al., 2024; Ray et al., 2024). ChatGPT is the most popular among the general public (Aydin & Karaaslan, 2023), while Bard is backed by Google's hegemonic position in the domain of algorithmic systems and search engines (Mager et al., 2023). The release of ChatGPT by OpenAI at the end of 2022 marked an AI milestone (Fui-Hoon Nah et al., 2023). Other companies have since started offering similar products based on so-called large language models (LLMs), which use the technology of generative pre-trained transformers (Hadi et al., 2023). Bard was released by Google in March 2023. While similar to ChatGPT, it is more focused on accurate answers (Singh et al., 2023).

The online applications of the two chatbots were fed three text queries related to the research objective. The first asked about use cases of AI for sustainability, categorized by industry. It did not ask for the use of any specific industrial classification. Some differences between these two tools were expected. Rolnick et al. (2023) reviewed the potential contributions of ML to climate change mitigation in the following main sectors: electricity systems, transportation, buildings and cities, industry, farms and forests, carbon dioxide removal, education, and finance. From their systematic literature review on the impacts of AI on sustainability, Kar et al. (2022) derived some use cases in the following industries: water management, agriculture, transportation, healthcare, manufacturing, construction, and financial services. The second query asked ChatGPT and Bard to provide use cases of AI for each SDG. This question echoed the approach of academic studies aimed at evaluating the impact of AI on individual SDGs (Vinuesa et al., 2020; Pigola et al., 2021; Bachmann et al., 2022). Finally, the third question asked about the main AI techniques that can help with sustainability. Based on the definition of AI given by McCarthy et al. (2006), it was expected that the techniques identified in response to this query would be related to the following: visual information, reasonable knowledge representation for decision making, automated mind-thinking to process stored knowledge, and ML to extract common patterns from the available information (Pigola et al., 2021).

4 Analysis of AI and Sustainability in the News

To gather the news for analysis, a search was conducted in Google News in English. The results were gathered by automatic download through the application programming interface (API). The search query was "artificial intelligence sustainability". The search returned 405 articles. After some cleaning of the data, 400 articles were retained.

The articles in the data set spanned a time frame running from November 2022 (when OpenAI released ChatGPT) to August 2023 (when the search was performed), thus allowing the analysis of a timespan of almost one year where the most recent advances and contributions from AI developments to SD could be framed. The main fields were Title, Body, Description, Date published, Source, and Link. The articles came from 132 sources. Inventiva.co.in from India and zawya.com from the Middle East were the two most common sources, with 68 and 54 articles, respectively. All other sources had 12 or fewer articles.

4.1 Text Analysis by Word Category

The first analysis of the data set consisted of looking for the most used words. The text in an article's title and body summarize the information contained in that article. It can therefore show the main words and ideas related to sustainability and AI in newspapers. Part-of-speech tagging was used instead of pure word counting to perform this analysis.[1] Part-of-speech tagging is a method used to classify words by their grammatical category. It reflects whether a word is a verb, a common noun, a proper noun, an adjective, or some other part of speech. Next, the words were ranked by number of occurrences. Both tasks were performed with Graphext. The categories used for the analysis were proper nouns (e.g., Google), nouns, and adjectives, respectively shown in columns 1, 2 and 3 of Fig. 1.

Among the proper nouns, AI is the most mentioned proper noun, with more than 200 mentions in the 400 articles. Other nouns mainly refer to places and companies, giving an idea of where to look for innovations. Interestingly, "United" and "States" are among the most mentioned proper nouns, as expected given that the United States is one of the countries with a strong tradition in AI research. The position of India in the first quartile is worthy of further investigation, given that the United States and India are separated by a considerable gap in technological literacy, access to digital equipment, and availability of suitable infrastructure (Bhutoria, 2022).

Among the common nouns, on the top are those directly related to the topic, followed by "industry", "company", "solution", and "business". This finding may imply that news stories talk more about the private sector than the public sector or academia, confirming the need to encourage the public sector to become more

[1] https://www.graphext.com/docs/analysis-types-text

involved in AI developments, not only as a regulator but also as a builder of AI capacity in public organizations (Wirtz & Müller, 2019) for purposes such as more efficient and higher-quality management of public services (Di Vaio et al., 2020). "Energy" is one of the few sectors that appears in the list of nouns, highlighting the importance of energy in relation to this topic. The term "energy" may also be related to one of the main environmental impacts of AI, namely the large amounts of energy required to process vast quantities of data (Wu, Raghavendra, et al., 2022).

Among the adjectives there are some that help to shape the topic such as "digital", because AI depends on data flows to learn and requires other DTs to make that data useful (Palomares et al., 2021; Pigola et al., 2021); "innovative", since innovation must be present in any "process" and "change" of the "industry" towards sustainability (Kuo & Smith, 2018); "environmental", that is one of the keys of "sustainability"; and "financial", "key" to the "effective" "development" of AI for SD (Al-Sartawi et al., 2022).

4.2 *Topic Clustering in News Description*

The keyword extraction and clustering analysis looks for the main keywords of the news in the description field. A clustering algorithm based on topic similarity was then used to classify the keywords.[2] The main goal of this analysis was to provide an overview of the topics in the data set, their representativeness, and the links between keywords.

The clustering tool names the clusters with the main keywords of the cluster. For example, the eighth cluster was initially named "big datum, come year, July". For analysis and clarity, the clusters have been renamed to show the main category of the keywords, where possible. Figure 2 shows the most used topics in the news by percentage. It offers an aggregate view of the topics discussed in the public debate.

The news about AI and SD firstly develop on where is this discussion happening (location), secondly on what are the actions taken, and thirdly on the industries involved. Notably, AI uses and industries are likely to be influenced by the location of most AI research (Vinuesa et al., 2020). There are two industry clusters. One is generic, and the other is for specific industries such as energy, agriculture, and waste management. This finding reflects the importance of these three industries for sustainability and AI (Goralski & Tan, 2020). Following industry, the most popular keywords refer to the entities involved. These entities are mainly countries, companies, or organizations, which must work cohesively to direct AI toward sustainable development (Di Vaio et al., 2020; Zhao et al., 2022). This idea is closely related to SDG 16 (Peace, justice, and strong institutions) and SDG 17 (Partnerships for the goals). Risks and benefits are next in the list. Unsurprisingly, AI elicits concerns about its dangers, whereas sustainability is commonly seen as beneficial (Goralski

[2] https://www.graphext.com/docs/analysis-types-text

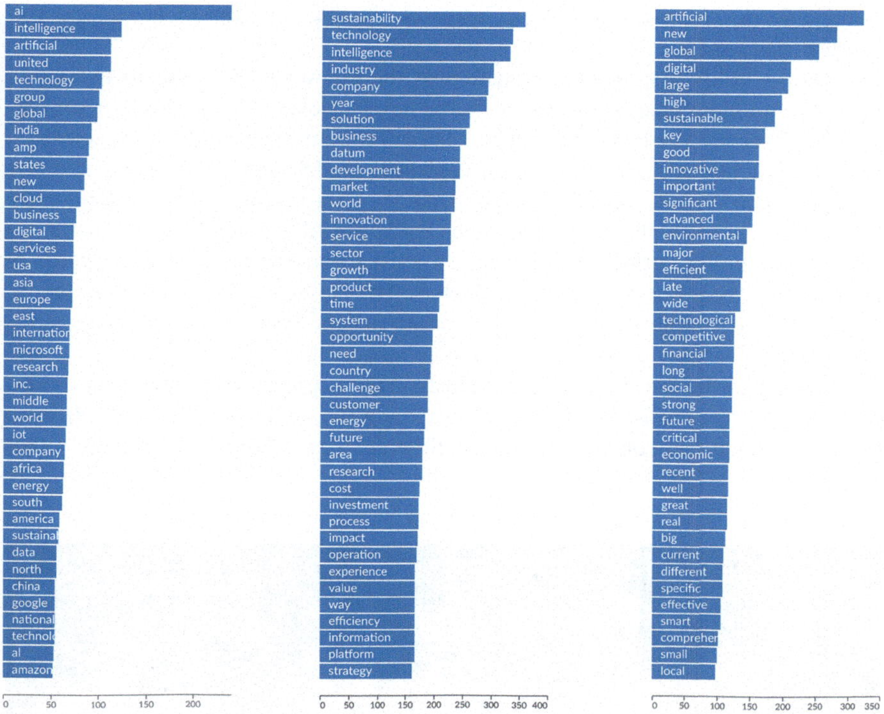

Fig. 1 Identification of proper names, nouns, and adjectives in news' titles with Graphext

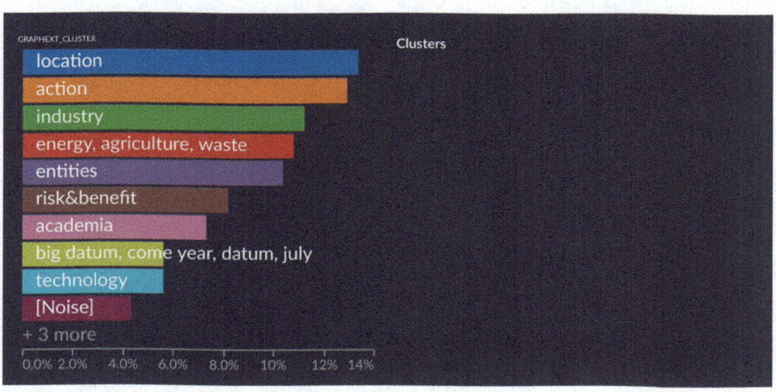

Fig. 2 Topic clusters generated with Graphext, by percentage of news in the dataset

& Tan, 2020). The seventh most common keyword cluster relates to academia, with topics such as university, research, and article, paying more attention to the ethical, social, and environmental controversies of the relationship between AI and sustainable development (Del Río Castro et al., 2021; Meitei et al., 2023; Wu, Shirkey, et al., 2022).

Figure 3 offers a spatial visualization of the keywords. A keyword extraction algorithm took the main keywords from the news articles. An algorithm of similarity then generated the links and distance between them to group them into clusters. Each dot is a keyword, but only some keywords are labeled in the figure. The distance between dots is measured by the similarity algorithm of the two keywords. The size of the dot is the number of times the keyword is mentioned. A cluster is a group with similar entities. In this case, a cluster consists of similar keywords in each article. The color denotes these clusters. The clusters in Fig. 3 are the same as those in Fig. 2. Purple keywords in the bottom of Fig. 3 are considered noise terms because the algorithm could not assign them to a cluster.

Focusing on the four more relevant clusters, the Location one is the most significant. The biggest dot is the "United States" location, indicating that it is the most important. Apart from other concrete locations such as "Europe", or "Africa", other

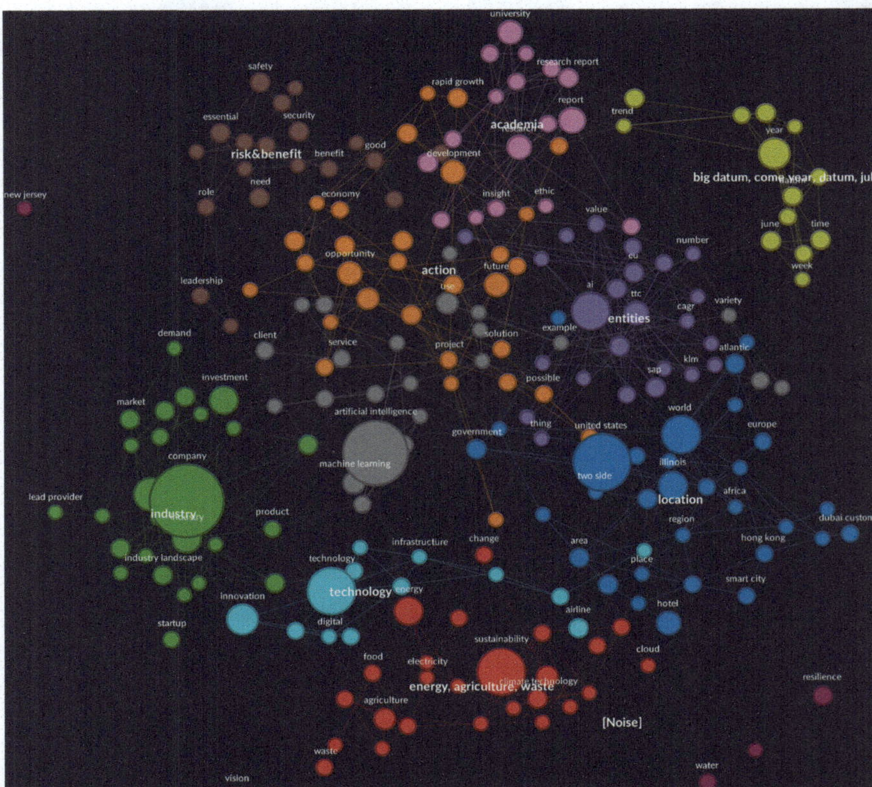

Fig. 3 Topic clustering in a graph-based chart, clustered by similarity with Graphext

generic locations refer to the application of AI technologies in specific contexts such as "home", "building", "hotel", and "city" (also "smart city").

The Action cluster involves keywords related to the transformative potential of AI, such as "opportunity", "solution", or "future". The Industry cluster gathers terms related to different types of private actors such as "company", "startup", or "lead provider" and terms related to economic system such as "economy", "market", "demand", or "investment". The Energy, agriculture, waste cluster is the most related is the most strongly related to sustainability and offers some interesting insights into the economic sectors that are most closely related to the topic. The main keyword is "sustainability", represented by the large dot in the center of the image. In the upper left corner, "energy" is the second biggest keyword, surrounded by other keywords such as "electricity" and "renewable". The bottom left corner contains keywords related to "agriculture", including "food waste" and "waste". On the right are keywords about the "environment" and "climate change". Examples include "carbon", "deforestation", and "planet".

4.3 Sentiment Analysis and Emotion Categorization

Finally, the titles were examined from a sentiment and emotion perspective. Sentiment analysis classifies short texts (in this case, article titles) into positive, negative, or neutral (Kiritchenko et al., 2014). Neutrality refers to neither negative nor positive. It is different from journalistic objectivity, sometimes called journalist neutrality, which means that stories are reported in an unbiased, even-handed, and impartial manner (Benham, 2020). The sentiment analysis was performed with the Cardiff NLP algorithm in the Graphext tool. The emotional classification offers a more nuanced view of emotions in the text. It is useful to see the public perception in the news on sustainability and AI.

From the 400 articles in the data set, 398 were analyzed by title using Cardiff NLP. The results are shown in Fig. 4. Most titles were either neutral (196) or positive (188). Only 14 titles were negative. The conclusion is that the use of AI for sustainability is seen mostly as non-negative, echoing the results reported in the academic research (Vinuesa et al., 2020).

The emotional analysis in Fig. 5 reveals that "excitement" was the main category, closely followed by "realization", both of which were positive. The category "caring" was used in reference to AI concerns. The rest of the categories were mainly positive. Examples include "curiosity", "approval", "desire", and "optimism". This analysis is of interest because AI generates serious concerns in public opinion. These concerns have reached such a level that some eminent figures in the tech world have advocated stopping research into AI until it is regulated given the risk it could pose to civilization (Yeh et al., 2021).

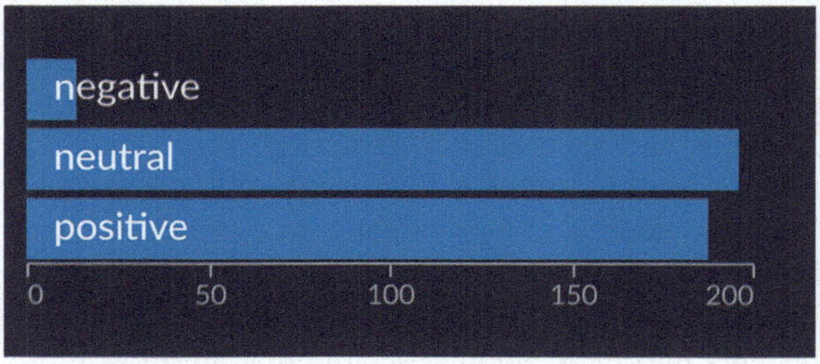

Fig. 4 Sentiment analysis with Graphext's Cardiff NLP algorithm

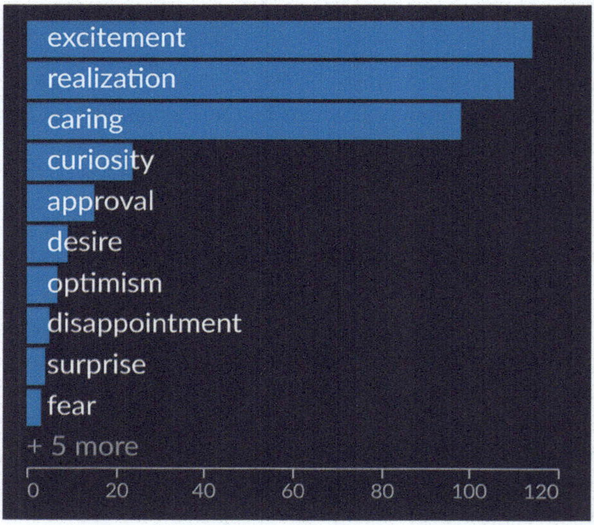

Fig. 5 Emotional analysis with Graphext's Cardiff NLP algorithm

5 AI and Sustainability Seen by AI (ChatGPT & Bard)

5.1 Use Cases from AI

To see how AI tools report on the intersection of AI and sustainable development, the first question was as follows: "Can you provide use cases of artificial intelligence (AI) for sustainability, categorized by industry?"

There were only a few differences in the industry classification given by the two chatbots. Notably, the financial and healthcare sectors were not identified by Bard. The industry classification provided by ChatGPT resembled the one used by Kar et al. (2022). Neither of the two chatbots referred to education or buildings and cities, which were cited by Rolnick et al. (2023). Both evidence the importance of the energy and agriculture sectors, as revealed by the analysis of Google News, but ChatGPT does not make a concrete reference to the waste industry. Nevertheless, some references to waste reduction are made across the other sectors in both chatbots responses.

Table SM1 (Supplementary Material) shows a comparison of the use cases proposed by the two chatbots for the same industry. We put the focus on the energy, agriculture and waste sectors, revealed as significant in the previous analyses, and where some relevant applications of AI would be providing fine-grained renewable energy forecasts for power companies, applying software/hardware tools for precision agriculture, or optimizing supply chains to reduce waste (Rolnick et al., 2023).

In the energy sector, the use cases proposed by Bard and ChatGPT included references to smart grids, renewable energy, and energy efficiency. In the renewable energy sector, Fan et al. (2023) highlight how AI and deep learning (DL) have been effectively utilized in optimizing energy management, fault detection, and power grid stability.

In the agriculture sector, the only common technique was precision agriculture/farming, based on applying AI to optimize crop yields and reduce the use of resources, supported by other DTs such as sensors (Vincent et al., 2019) or drones (Spanaki et al., 2022). Both Bard and ChatGPT also refer to supply chain optimization to reduce food waste in this sector. Bard highlights AI applications in sustainable aquaculture to monitor water quality, or to identify and prevent diseases. ChatGPT refers to AI algorithms for the early identification of crops diseases.

In the waste sector, Bard highlights the contributions of AI to improve waste collection, disposal, and recycling, by optimizing routes and predicting where waste will be generated, and identifying the most sustainable ways to dispose of different types of waste. Since waste reduction is also related to sustainable production and consumption, Pigola et al. (2021) refer to AI applications in production planning adapted to predicted consumption patterns to avoid unnecessary waste.

5.2 *How Can AI Help with UN 17 SDGs?*

The second question was posed as follows: "Can you provide use cases of artificial intelligence (AI) for sustainability for each UN Sustainable Development Goal?"

Table SM2 (Supplementary Material) shows that Bard offered a greater variety of use cases for each SDG. In general, ChatGPT focused on the application of ML to predict or identify opportunities or risks related to each SDG (ML is mentioned in all the SDGs except SDG 4 and SDG 5). In contrast, Bard barely mentioned any predictive use of AI techniques and focused on AI developments directed at providing new technologies to serve the different SDGs. Moreover, in SDGs 5 and 10 to 17, Bard highlighted the importance of AI to raise public awareness around each one of those SDGs. Finally, both ChatGPT and Bard mentioned the use of AI to track and monitor indicators related to each SDG.

The AI for SDGs Observatory offers in its website[3] a list of projects categorized by SDG. Cowls et al. (2021) conclude that these are an increasingly global

[3] https://ai-for-sdgs.academy/casebase

phenomenon but not equally distributed across the SDGs. With the purpose of enquiring the two chatbots about these projects, the following sub-question was used for each of the SDGs: "Can you provide some examples of company or project that is applying AI for the achievement of SDG x?" Table 1 offers one such example of project for each one of the 17 SDGs. Table SM3 of the Supplementary Material offers the complete lists of projects reported by each chatbot.

5.3 AI Techniques for Sustainability

Finally, the third question posed to both Bard and ChatGPT was as follows: "What are the main AI techniques that can help in Sustainability?"

Both chatbots identified machine learning (ML), deep learning (DL), natural language processing (NLP), optimization, and simulation and modeling techniques. ChatGPT also referred to data analytics (data mining would be included in Bard's NLP category and predictive analytics in Bard's ML category), as well as image and video analysis (included in Bard's DL category). Bard also included reinforcement learning (RL) in its list (included in ChatGPT's ML category). Table SM4 (Supplementary Material) shows a comparison of the AI techniques proposed by the two chatbots.

In the five techniques proposed by both chatbots, Bard focused on explaining specific uses, whereas ChatGPT seemed to offer a list of techniques with a brief description of their applications. Nonetheless, there were no great differences in the tools and applications identified in each response. For instance, regarding the description of ML, ChatGPT, refers to three specific types of learning (Supervised, Unsupervised and Reinforcement Learning) and offers some examples of their application, while Bard first offers a general definition of ML and then provides three examples similar to those of ChatGPT.

Adding to the vast array of scientific literature dealing with the relation between AI and sustainability, mainly guided by a positive view, this chapter has turned the look towards a different knowledge base considering both online news and chatbots responses. The findings confirm the positive bias in the presentation of the AI and SD intersection and allow to derive some relevant insights for several different audiences: 1) researchers should put a bigger focus on the potential conflicts between AI and SDGs; 2) these are relevant areas where policy makers should increase their regulatory pressure; and 3) corporate leaders, entrepreneurs, and investors should not take for granted the positive contribution of AI to SD, making a balanced analysis of its impact on both SD and the bottom line.

Table 1 Examples of projects or companies contributing to each SDG through AI applications, based on chatbots repsonses

SDG1: No poverty	Operating in Chile and Mexico, Destacame utilizes AI to assess credit risk for individuals without a credit history. Through alternative data analysis, such as utility bill payments and mobile phone records, they generate credit scores, enabling those previously excluded from financial services to access loans and improve their economic situation.
SDG2: Zero Hunger	Farmwise integrates AI and robotics to enhance farming practices. Their machines use computer vision to identify weeds and other unwanted plants, allowing for precise and efficient removal without harming crops. This helps improve crop yields and contributes to sustainable agriculture, ultimately aiding in addressing hunger.
SDG3: Good Health and Well-being	Google's DeepMind Health collaborates with healthcare providers to apply AI for various purposes, such as analyzing medical images, predicting patient deterioration, and assisting in disease diagnosis. For instance, they've developed algorithms to assist with eye disease diagnosis and to interpret complex medical imaging.
SDG4: Quality Education	IBM's Watson Education applies AI to personalize learning experiences. By analyzing students' strengths, weaknesses, and learning patterns, it offers educators insights to create customized learning plans and interventions, aiming to improve learning outcomes.
SDG5: Gender Equality	HireVue uses AI in recruitment processes to minimize biases in hiring. By employing machine learning algorithms to assess job applicants based on skills and qualifications rather than demographic factors, it aims to create fairer opportunities for all candidates, including women.
SDG6: Clean Water and Sanitation	Drinkwell uses AI-driven water purification systems to provide clean drinking water in communities facing water contamination issues. Their systems remove contaminants and pathogens, ensuring safe drinking water in areas where clean water access is limited.
SDG7: Affordable and Clean Energy	Bidgely provides AI-powered solutions for energy disaggregation, which helps consumers understand their energy usage at a granular level. By analyzing data from smart meters, Bidgely offers personalized insights to encourage energy efficiency and reduce overall consumption.
SDG8: Decent Work and Economic Growth	ManpowerGroup employs AI in workforce management solutions. Their AI-driven platforms analyze labor market trends, skill gaps, and workforce demands to provide insights that assist companies in making informed decisions about hiring, training, and workforce planning.
SDG9: Industry, Innovation and Infrastructure	Siemens employs AI in various industrial applications to enhance manufacturing processes. They use AI for predictive maintenance in machinery, optimizing production schedules, and improving overall efficiency in industrial settings.
SDG10: Reduced Inequalities	Gapminder uses AI-powered data visualization tools to highlight global inequalities. By presenting data in accessible ways, they raise awareness about disparities in income, health, education, and other factors, fostering informed discussions about reducing inequalities.
SDG11: Sustainable Cities and Communities	StreetLight Data: StreetLight Data utilizes AI-driven analytics to provide insights into urban mobility patterns. By analyzing traffic and movement data, they help city planners make informed decisions about transportation infrastructure and traffic management.

(continued)

Table 1 (continued)

SDG12: Responsible Consumption and Production	Too Good To Go employs AI to reduce food waste by connecting consumers with surplus food from restaurants and grocery stores. Their app allows users to purchase leftover food at discounted prices, preventing it from being discarded.
SDG13: Climate Action	CarbonCure utilizes AI and data analytics to reduce carbon emissions in the construction industry. Their technology injects captured CO2 into concrete, sequestering carbon and reducing the carbon footprint of construction projects.
SDG14: Life Below Water	Open-Oceans Global applies AI to analyze large sets of oceanic data. Their platform uses machine learning algorithms to assess the health of marine ecosystems and predict the impact of human activities on ocean environments.
SDG15: Life on Land	NatureServe applies AI to ecological data to assess and monitor biodiversity. Their platform uses machine learning to analyze species data, habitat information, and conservation statuses, aiding in land management decisions to protect biodiversity.
SDG16: Peace, Justice and Strong Institutions	The United Nations Global Pulse utilizes AI and data analytics to support the United Nations in addressing societal challenges. They analyze big data to generate insights that inform policy-making and foster resilience, contributing to stronger institutions for sustainable development.
SDG17: Partnerships for the Goals	DataKind collaborates with nonprofit organizations and governments to apply AI and data science for social good. They foster partnerships between data scientists and mission-driven organizations to address various societal challenges.

6 Future

Despite the debate around the positive and negative aspects of AI for sustainability, there is reasonable doubt as to the extent to which AI can contribute to sustainable development when the system itself is "inherently unsustainable" (Falk & van Wynsberghe, 2023). To the extent that AI is part of a complete sociotechnical system, it cannot be analyzed as a simple tool isolated from its context and controlled by only a few actors, companies, and countries under a strong capitalist system (Sætra, 2021). Considerations of inclusivity, sustainability, universality, inequality, or equity make it harder to see AI as an enabler of many goals (Sætra, 2021).

Moreover, some studies supporting the idea that AI can potentially enable sustainability have been conducted in controlled laboratory environments, based on limited data sets or prototypes (Vinuesa et al., 2020). Sætra (2021) has warned that the limited successes obtained in closed experimental settings are not necessarily transferable to the open systems of real life. It has also been noted that AI can do little to confront "natural stupidity" (Vuong et al., 2019), which often translates into excessive bureaucracy that slows down the adoption and use of the contributions of AI or into a lack of sociopolitical commitment and financial support.

Thus, future research should pay more attention to the ethical, social and environmental controversies of the relationship between AI and SD (del Río Castro et al., 2021; Meitei et al., 2023; Wu, Shirkey, et al., 2022). It should independently and rigorously seek to analyze the ethics and sustainability of AI (Sætra, 2021). It should be careful to avoid funding and commercial bias, as signaled by Vinuesa et al. (2020), which can prevent AI projects with the potential to achieve certain SDGs from being supported if alternative projects show better chances of maximizing profit. Research on the possibilities of developing and directing AI toward the common good should integrate sociotechnical concerns and evaluate possible third-order effects (Ligozat et al., 2022). The design and development of AI solutions should be promoted in less wealthy nations and regions (Vinuesa et al., 2020). The interrelations among the SDGs should be acknowledged, considering that AI can help achieve some of them at the cost of others and that a biased development of AI can increase inequalities between and within countries. An interdisciplinary approach is thus imperative (Tomašev et al., 2020). Transparency and safety are also important aspects to account for (Vinuesa et al., 2020), and more profound research in places and communities is required (Meitei et al., 2023).

Finally, the role of governments is essential to facilitate a suitable mix of strategic actions, policy, and finance (Walsh et al., 2020). Vinuesa et al. (2020, p 6) reported that "regulatory oversight should be preceded by regulatory insight." Policymakers should work with AI experts, build AI capabilities in relevant public sector bodies, and release data that may be relevant to the achievement of the SDGs (Rolnick et al., 2023). Truby (2020) focused on regulatory approaches to minimize the negative impact of AI on SDGs, based on ensuring the development of verifiable, accountable, and transparent AI systems and refusing to adopt AI technologies that do not comply with SDG principles. When integrating AI in public decision making, governments have also to ensure that data biases in models do not result in biased policy implementation (Bachmann et al., 2022). Legislation on AI transparency and accountability and on privacy and data ownership issues is needed, as well as the adoption of ethical standards are also needed to develop a sustainable-development-friendly AI (Vinuesa et al., 2020). Governments should encourage the development and implementation of methodologies for sustainability impact assessments of AI technologies and provide private sector actors with incentives to develop and integrate AI applications aligned with the SDGs (Gailhofer et al., 2021). AI regulations also require a continual effort of risk identification to avoid potential negative effects on social cohesion, environmental conservation, democratic principles, and human rights (Fernández-Aller et al., 2021). Continuous surveillance of AI developments and impact on sustainability should be supported by the creation of interdisciplinary oversight bodies and platform or social media councils (Mager et al., 2023).

In summary, to ensure the responsible development and application of AI technologies, governments must intervene through regulatory standards, physical infrastructures, and digital systems (Sachs et al., 2019). They must also encourage knowledge management and sharing tools capable of promoting dialogue between humans and machines (Di Vaio et al., 2020). In this chapter, the interaction with the Bard and ChatGPT chatbots and the application of Graphext to the analysis of online news verifies that the contribution of AI to sustainable development can be invaluable, if it is primarily driven by human intelligence.

References

Allen, C., Metternicht, G., & Wiedmann, T. (2018). Initial progress in implementing the Sustainable Development Goals (SDGs): A review of evidence from countries. *Sustainability Science, 13*, 1453–1467. https://doi.org/10.1007/s11625-018-0572-3

Al-Sartawi, A. M. A. M., Hussainey, K., & Razzaque, A. (2022). The role of artificial intelligence in sustainable finance. *Journal of Sustainable Finance & Investment*. (Early Access). https://doi.org/10.1080/20430795.2022.2057405

Aydin, Ö., & Karaaslan, E. (2023). Is ChatGPT leading generative AI? What is beyond expectations? *Academic Platform Journal of Engineering and Smart Systems, 11*(3), 118–134. https://doi.org/10.21541/apjess.1293702

Bachmann, N., Tripathi, S., Brunner, M., & Jodlbauer, H. (2022). The contribution of data-driven technologies in achieving the sustainable development goals. *Sustainability, 14*(5), 2497. https://doi.org/10.3390/su14052497

Benham, J. (2020). Best practices for journalistic balance: Gatekeeping, imbalance and the fake news era. *Journalism Practice, 14*(7), 791–811. https://doi.org/10.1080/17512786.2019.1658538

Bhutoria, A. (2022). Personalized education and artificial intelligence in the United States, China, and India: A systematic review using a human-in-the-loop model. *Computers and Education: Artificial Intelligence, 3*, 100068. https://doi.org/10.1016/j.caeai.2022.100068

Chui, M., Harryson, M., Valley, S., Manyika, J., & Roberts, R. (2018). *Notes from the AI frontier applying AI for social good*. Discussion paper. McKinsey Global Institute.

Cowls, J., Tsamados, A., Taddeo, M., & Floridi, L. (2021). A definition, benchmark and database of AI for social good initiatives. *Nature Machine Intelligence, 3*(2), 111–115. https://doi.org/10.1038/s42256-021-00296-0

Del Río Castro, G., González-Fernández, M. C., & Uruburu-Colsa, A. (2021). Unleashing the convergence amid digitalization and sustainability towards pursuing the Sustainable Development Goals (SDGs): A holistic review. *Journal of Cleaner Production, 280*(1), 122204. https://doi.org/10.1016/j.jclepro.2020.122204

Denning, P. J., & Denning, D. E. (2020). Dilemmas of artificial intelligence. *Communications of the ACM, 63*(3), 22–24. https://doi.org/10.1145/3379920

Di Vaio, A., Palladino, R., Hassan, R., & Escobar, O. (2020). Artificial intelligence and business models in the sustainable development goals perspective: A systematic literature review. *Journal of Business Research, 121*, 283–314. https://doi.org/10.1016/j.jbusres.2020.08.019

Falk, S., & van Wynsberghe, A. (2023). Challenging AI for sustainability: What ought it mean? *AI and Ethics*, 1–11. https://doi.org/10.1007/s43681-023-00323-3

Fan, Z., Yan, Z., & Wen, S. (2023). Deep learning and artificial intelligence in sustainability: A review of SDGs, renewable energy, and environmental health. *Sustainability, 15*(18), 13493. https://doi.org/10.3390/su151813493

Fernández-Aller, C., de Velasco, A. F., Manjarrés, Á., Pastor-Escuredo, D., Pickin, S., Criado, J. S., & Ausín, T. (2021). An inclusive and sustainable artificial intelligence strategy for europe based on human rights. *IEEE Technology and Society Magazine, 40*(1), 46–54. https://doi.org/10.1109/MTS.2021.3056283

Frijters, P., & Velamuri, M. (2010). Is the internet bad news? The online news era and the market for high-quality news. *Review of Network Economics, 9*(2). https://doi.org/10.2202/1446-9022.1187

Fui-Hoon Nah, F., Zheng, R., Cai, J., Siau, K., & Chen, L. (2023). Generative AI and ChatGPT: Applications, challenges, and AI-human collaboration. *Journal of Information Technology Case and Application Research, 25*(3), 277–304. https://doi.org/10.1080/15228053.2023.2233814

Fuso Nerini, F., Sovacool, B., Hughes, N., Cozzi, L., Cosgrave, E., Howells, M., Tavoni, M., Tomei, J., Zerriffi, H., & Milligan, B. (2019). Connecting climate action with other Sustainable Development Goals. *Nature Sustainability, 2*(8), 674–680. https://doi.org/10.1038/s41893-019-0334-y

Gailhofer, P., Herold, A., Schemmel, J. P., Scherf, C. U., Köhler, A. R., & Braungardt, S. (2021). *The role of artificial intelligence in the European green deal*. Technical Report, Study for the Special Committee on Artificial Intelligence in a Digital Age (AIDA), Policy Department for Economic, Scientific and Quality of Life Policies, European Parliament. Accessed January 9, 2024, from https://www.europarl.europa.eu/RegData/etudes/STUD/2021/662906/IPOL_STU(2021)662906_EN.pdf

Gan, R. K., Ogbodo, J. C., Wee, Y. Z., Gan, A. Z., & Arcos-González, P. (2024). Performance of Google bard and ChatGPT in mass casualty incidents triage. *The American Journal of Emergency Medicine, 75*, 72–78. https://doi.org/10.1016/j.ajem.2023.10.034

Goralski, M. A., & Tan, T. K. (2020). Artificial intelligence and sustainable development. *The International Journal of Management Education, 18*(1), 100330. https://doi.org/10.1016/j.ijme.2019.100330

Hadi, M. U., Tashi, Q. A., Qureshi, R., Shah, A., Muneer, A., Irfan, M., Zafar, A., Shaikh, M. B., Akhtar, N., Wu, J., & Mirjalili, S. (2023). Large language models: A comprehensive survey of its applications. *Challenges, Limitations, and Future Prospects*. https://doi.org/10.36227/techrxiv.23589741.v2

Halkos, G., & Gkampoura, E. C. (2021). Where do we stand on the 17 Sustainable Development Goals? An overview on progress. *Economic Analysis and Policy, 70*, 94–122. https://doi.org/10.1016/j.eap.2021.02.001

International Telecommunication Union. (2023). *AI4Good global summit*. Retrieved November 25, 2023, from https://aiforgood.itu.int

Kar, A. C., Choudhary, S. K., & Singh, V. K. (2022). How can artificial intelligence impact sustainability: A systematic literature review. *Journal of Cleaner Production, 376*, 134120. https://doi.org/10.1016/j.jclepro.2022.134120

Khakurel, J., Penzenstadler, B., Porras, J., Knutas, A., & Zhang, W. (2018). The rise of artificial intelligence under the lens of sustainability. *Technologies, 6*(4), 100. https://doi.org/10.3390/technologies6040100

Kiritchenko, S., Zhu, X., & Mohammad, S. M. (2014). Sentiment analysis of short informal texts. *Journal of Artificial Intelligence Research, 50*, 723–762. https://doi.org/10.1613/jair.4272

Kuo, T. C., & Smith, S. (2018). A systematic review of technologies involving eco-innovation for enterprises moving towards sustainability. *Journal of Cleaner Production, 192*, 207–220. https://doi.org/10.1016/j.jclepro.2018.04.212

Ligozat, A. L., Lefevre, J., Bugeau, A., & Combaz, J. (2022). Unraveling the hidden environmental impacts of AI solutions for environment life cycle assessment of AI solutions. *Sustainability, 14*(9), 5172. https://doi.org/10.3390/su14095172

Lu, S., & Serafeim, G. (2023, June 12). How AI will accelerate the circular economy. *Harvard Business Review*. https://hbr.org/2023/06/how-ai-will-accelerate-the-circular-economy

Mager, A., Norocel, O. C., & Rogers, R. (2023). Advancing search engine studies: The evolution of Google critique and intervention. *Big Data & Society, 10*(2), Editorial. https://doi.org/10.1177/20539517231191528

McCarthy, J., Minsky, M. L., Rochester, N., & Shannon, C. E. (2006). A proposal for the Dartmouth summer research project on artificial intelligence, August 31, 1955. *AI Magazine, 27*(4), 12. https://doi.org/10.1609/aimag.v27i4.1904

Meitei, A. J., Rai, P., & Rajkishan, S. S. (2023). Application of AI/ML techniques in achieving SDGs: A bibliometric study. *Environment, Development and Sustainability*, 1–37. https://doi.org/10.1007/s10668-023-03935-1

Nghiem, L. T. P., Papworth, S. K., Lim, F. K. S., & Carrasco, L. R. (2016). Analysis of the capacity of Google trends to measure interest in conservation topics and the role of online news. *PLoS One, 11*(3), e0152802. https://doi.org/10.1371/journal.pone.0152802

Nieto-Rodriguez, A. & Viana Vargas, R. (2023, October 27). The opportunities at the intersection of AI, sustainability, and project management. *Harvard Business Review*. https://hbr.org/2023/10/the-opportunities-at-the-intersection-of-ai-sustainability-and-project-management

Núñez-Merino, M., Maqueira-Marín, J. M., Moyano-Fuentes, J., & Castaño-Moraga, C. A. (2022). Industry 4.0 and supply chain. A systematic science mapping analysis. *Technological Forecasting and Social Change, 181*, 121788. https://doi.org/10.1016/j.techfore.2022.121788

Palomares, I., Martínez-Cámara, E., Montes, R., García-Moral, P., Chiachio, M., Chiachio, J., Alonso, S., Melero, F. J., Molina, D., Fernández, B., Moral, C., Marchena, R., de Vargas, J. P., & Herrera, F. (2021). A panoramic view and swot analysis of artificial intelligence for achieving the sustainable development goals by 2030: Progress and prospects. *Applied Intelligence, 51*(9), 6497–6527. https://doi.org/10.1007/s10489-021-02264-y

Patankar, A. A., & Bose, J. (2016). Bias based navigation for news articles and media. In *Natural language processing and information systems: 21st International Conference on Applications of Natural Language to Information Systems, NLDB 2016, Salford, UK, June 22–24, 2016, Proceedings 21* (pp. 465–470). Springer International Publishing. https://doi.org/10.1007/978-3-319-41754-7_50.

Peng, Y., Ahmad, S. F., Ahmad, A. Y. B., Al Shaikh, M. S., Daoud, M. K., & Alhamdi, F. M. H. (2023). Riding the waves of artificial intelligence in advancing accounting and its implications for sustainable development goals. *Sustainability, 15*(19), 14165. https://doi.org/10.3390/su151914165

Pigola, A., da Costa, P. R., Carvalho, L. C., Silva, L. F. D., Kniess, C. T., & Maccari, E. A. (2021). Artificial intelligence-driven digital technologies to the implementation of the sustainable development goals: A perspective from Brazil and Portugal. *Sustainability, 13*(24), 13669. https://doi.org/10.3390/su132413669

Purvis, B., Mao, Y., & Robinson, D. (2018). Three pillars of sustainability: In search of conceptual origins. *Sustainable Science, 14*(3), 681–695. https://doi.org/10.1007/s11625-018-0627-5

Ray, S. S., Peddinti, P. R. T., Verma, R. K., Puppala, H., Kim, B., Singh, A., & Kwon, Y. N. (2024). Leveraging ChatGPT and Bard: What does it convey for water treatment/desalination and harvesting sectors? *Desalination, 570*, 117085. https://doi.org/10.1016/j.desal.2023.117085

Rolnick, D., Donti, P. L., Kaack, L. H., Kochanski, K., Lacoste, A., Sankaran, K., Ross, A. S., Milojevic-Dupont, N., Jaques, N., Waldman-Brown, A., Luccioni, A. S., Maharaj, T., Sherwin, E. D., Mukkavilli, S. K., Kording, K. P., Gomes, C. P., Ng, A. Y., Hassabis, D., Platt, J. C., et al. (2023). Tackling climate change with machine learning. *ACM Computing Surveys, 55*(2), 1–96. https://doi.org/10.1145/3485128

Rudolph, J., Tan, S., & Tan, S. (2023). War of the chatbots: Bard, Bing Chat, ChatGPT, Ernie and beyond. The new AI gold rush and its impact on higher education. *Journal of Applied Learning & Teaching, 6*(1), 364–389. https://doi.org/10.37074/jalt.2023.6.1.23

Sachs, J. D., Schmidt-Traub, G., Mazzucato, M., Messner, D., Nakicenovic, N., & Rockström, J. (2019). Six transformations to achieve the sustainable development goals. *Nature Sustainability, 2*(9), 805–814. https://doi.org/10.1038/s41893-019-0352-9

Sætra, H. S. (2021). AI in context and the sustainable development goals: Factoring in the unsustainability of the sociotechnical system. *Sustainability, 13*(4), 1738. https://doi.org/10.3390/su13041738

Scholz, R., Bartelsman, E., Diefenbach, S., Franke, L., Grunwald, A., Helbing, D., Hill, R., Hilty, L., Höjer, M., Klauser, S., Montag, C., Parycek, P., Prote, J., Renn, O., Reichel, A., Schuh, G., Steiner, G., & Viale Pereira, G. (2018). Unintended side effects of the digital transition: European Scientists' messages from a proposition-based expert round table. *Sustainability, 10*(6), 2001. https://doi.org/10.3390/su10062001

Schwartz, R., Dodge, J., Smith, N. A., & Etzioni, O. (2020). Green AI. *Communications of the ACM, 63*, 54–63. https://doi.org/10.1145/3381831

Singh, S. K., Kumar, S., & Mehra, P. S. (2023, June). Chat GPT & Google Bard AI: A Review. In *2023 International Conference on IoT, Communication and Automation Technology (ICICAT)* (pp. 1–6). IEEE. https://doi.org/10.1109/ICICAT57735.2023.10263706.

Spanaki, K., Karafili, E., Sivarajah, U., Despoudi, S., & Irani, Z. (2022). Artificial intelligence and food security: Swarm intelligence of AgriTech drones for smart AgriFood operations.

Production Planning & Control, 33(16), 1498–1516. https://doi.org/10.1080/09537287.2021.1882688

Tomašev, N., Cornebise, J., Hutter, F., Mohamed, S., Picciariello, A., Connelly, B., et al. (2020). AI for social good: Unlocking the opportunity for positive impact. *Nature Communications, 11*(1), 2468. https://doi.org/10.1038/s41467-020-15871-z

Truby, J. (2020). Governing artificial intelligence to benefit the UN sustainable development goals. *Sustainable Development, 28*(4), 946–959. https://doi.org/10.1002/sd.2048

UN. (2015). *Transforming our world: The 2030 agenda for sustainable development.* United Nations, Department of Economic and Social Affairs.

Vermeer, S., Trilling, D., Kruikemeier, S., & de Vreese, C. (2020). Online news user journeys: The role of social media, news websites, and topics. *Digital Journalism, 8*(9), 1114–1141. https://doi.org/10.1080/21670811.2020.1767509

Vincent, D. R., Deepa, N., Elavarasan, D., Srinivasan, K., Chauhdary, S. H., & Iwendi, C. (2019). Sensors driven AI-based agriculture recommendation model for assessing land suitability. *Sensors, 19*(17), 3667. https://doi.org/10.3390/s19173667

Vinuesa, R., Azizpour, H., Leite, I., Balaam, M., Dignum, V., Domisch, S., Felländer, A., Langhans, S. D., Tegmark, M., & Fuso Nerini, F. (2020). The role of artificial intelligence in achieving the sustainable development goals. *Nature Communications, 11*(1). https://doi.org/10.1038/s41467-019-14108-y

Vuong, Q.-H., Ho, M.-T., Vuong, T.-T., La, V.-P., Ho, M.-T., Nghiem, K.-C. P., Tran, B. X., Giang, H.-H., Giang, T.-V., Latkin, C., Nguyen, H.-K. T., Ho, C. S. H., & Ho, R. C. M. (2019). Artificial intelligence vs. natural stupidity: Evaluating AI readiness for the Vietnamese medical information system. *Journal of Clinical Medicine, 8*(2), 168. https://doi.org/10.3390/jcm8020168

Walsh, P., Murphy, E., & Horan, D. (2020). The role of science, technology and innovation in the UN 2030 agenda. *Technological Forecasting and Social Change, 154*, 119957. https://doi.org/10.1016/j.techfore.2020.119957

WCED. (1987). *Our common future*. Oxford University Press.

Wirtz, B. W., & Müller, W. M. (2019). An integrated artificial intelligence framework for public management. *Public Management Review, 21*(7), 1076–1100. https://doi.org/10.1080/14719037.2018.1549268

Wu, C. J., Raghavendra, R., Gupta, U., Acun, B., Ardalani, N., Maeng, K., Chang, G., Aga, F., Huang, J., Bai, C., & Gschwind, M. (2022). Sustainable AI: Environmental implications, challenges and opportunities. *Proceedings of Machine Learning and Systems, 4*, 795–813. https://doi.org/10.48550/arXiv.2111.00364

Wu, S. R., Shirkey, G., Celik, I., Shao, C., & Chen, J. (2022). A review on the adoption of AI, BC, and IoT in sustainability research. *Sustainability, 14*(13), 7851. https://doi.org/10.3390/su14137851

Yeh, S. C., Wu, A. W., Yu, H. C., Wu, H. C., Kuo, Y. P., & Chen, P. X. (2021). Public perception of artificial intelligence and its connections to the sustainable development goals. *Sustainability, 13*(16), 9165. https://doi.org/10.3390/su13169165

Zhao, W., Yin, C., Hua, T., Meadows, M. E., Li, Y., Liu, Y., Cherubini, F., Pereira, P., & Fu, B. (2022). Achieving the Sustainable Development Goals in the post-pandemic era. *Humanities and Social Sciences Communications, 9*(1). https://doi.org/10.1057/s41599-022-01283-5

Artificial Intelligence Usefulness Effect on Business Performance with Trust

Samet Batuhan Güven, Gulin İdil S. Bolatan, and Tugrul Daim

Abstract The objective of this research is to examine the impact of AI usage on business performance in organizations. It focuses on assessing trust in AI, perceived usefulness of AI, and the ability of AI-enabled organizations to evaluate business performance. An online survey with 195 middle and senior executives, and engineers was conducted, and data was analyzed using SPSS AMOS software. Three hypotheses were analyzed: trust in AI's positive relationship with job performance, perceived usefulness of AI's positive link with trust in AI, and perceived usefulness of AI's positive association with job performance. Results support the second and third hypotheses, but not the first. This study enhances understanding of AI's impact on business performance. It provides insights into businesses' attitudes towards AI technologies and their effects on organizational performance by examining trust in AI, perceived usefulness of AI, and the ability to evaluate business performance. The findings highlight the positive influence of perceived usefulness of AI and trust in AI on business performance. The study emphasizes the importance of adopting and implementing AI technologies to enhance business performance. AI's capabilities, such as data analysis, prediction, learning, and automated decision-making, optimize operations and achieve faster, more efficient outcomes. Measuring business performance when utilizing AI is significant. AI-based predictive models improve demand forecasting, inventory management, customer experience, and personalization, leading to cost savings, time efficiency, and enhanced quality. Overall, this research offers valuable insights for managers and decision-makers, guiding the adoption and effective utilization of AI technologies in organizations.

S. B. Güven
Management Engineering Department, Alanya Alaaddin Keykubat University, Alanya, Turkey

G. İ. S. Bolatan
Industrial Engineering Department, Alanya Alaaddin Keykubat University, Alanya, Turkey

T. Daim (✉)
Mark O. Hatfield Cybersecurity & Cyber Defense Policy Center, Portland State University, Portland, OR, USA
e-mail: tugrul.u.daim@pdx.edu

M. T. Del Val Núñez et al. (eds.), *Artificial Intelligence and Business Transformation*, Contributions to Management Science,
https://doi.org/10.1007/978-3-031-58704-7_5

Understanding AI's role in business fosters research and development efforts, contributing to future innovations in the field.

1 Introduction

In the current business landscape, technological advancements and the process of digital transformation present businesses with new opportunities. Among these opportunities, artificial intelligence (AI) technologies offer various ways for businesses to gain a competitive edge and enhance their performance. With its ability to analyze complex data, make predictions, learn, and automate decision-making, AI has the potential to optimize operational processes and deliver faster and more efficient outcomes. As a result, businesses are increasingly adopting and implementing AI technologies.

This research aims to explore the impact of AI adoption on business performance. Specifically, it will focus on examining trust in AI, the usefulness of AI, and measuring the business performance of organizations utilizing AI. This study provides a valuable opportunity to gain insights into enterprises' attitudes towards AI technologies and their effects on business performance.

According to La Torre et al. (2021), trust and transparency pose potential risks to the efficacy of AI. AI applications have the potential to enhance the scope, depth, and flexibility of stakeholder interactions. However, they also present potential risks to the trust, precision, and safety of interactions, particularly those involving innovations (Jiang et al., 2023). Trust is a critical factor for knowledge management (Ozpamuk et al., 2023). On the other hand, the transformative character of AI and its capacity to induce significant alterations in business processes is widely recognized and is expected to escalate in the upcoming years (Jorzik et al., 2023). AI enhances business performance (Cammarano et al., 2023; Shen et al., 2023).

The primary objective of this research is to uncover the influence of AI adoption on business performance within organizations. Additionally, we will assess the usefulness of AI and the impact of relying on AI on business performance. Moreover, we will investigate whether AI creates tangible value for businesses by examining the ability of AI-utilizing organizations to measure their business performance.

2 Literature Review

This study investigates the impact of trust in artificial intelligence, the benefits of artificial intelligence, and business performance. Primarily, a review of the relevant literature was conducted, focusing on the identification of conceptualized relationships among the research variables.

2.1 Artificial Intelligence

Different definitions of artificial intelligence (AI) have been developed, with researchers offering varying perspectives. Wang et al. (2019) define AI as the intelligent behavior displayed by machines, while Demlehner and Laumer (2020) describe it as a computer system that can perceive, learn, judge, or plan without explicit programming.

Understanding AI requires distinguishing between the terms "artificial" and "intelligence." "Artificial" refers to human-made things rather than natural creations (Mikalef & Gupta, 2021), while "intelligence" describes cognitive processes such as learning and reasoning. By combining these concepts, AI is defined as creating methodologies and algorithms that allow computer programs to analyze data, identify trends, solve problems, draw conclusions, and communicate with people.

AI acts as an intelligent agent, replicating human abilities by making decisions based on environmental inputs (Eriksson et al., 2020). Its primary objective is to mimic human cognition, including learning and information processing. AI is often referred to as cognitive technology, enabling computers to think and behave more like humans (Bytniewski et al., 2020). Some emphasize the importance of AI carrying out intelligent activities without explicit programming (Demlehner & Laumer, 2020), decoding accurate external data, learning from it, and accomplishing goals using flexible methods (Makarius et al., 2020), without rigidly following established rules or sequences of operations (Demlehner & Laumer, 2020).

Two primary methods define AI. The first considers AI as a tool to complete difficult or time-consuming tasks (Demlehner & Laumer, 2020; Makarius et al., 2020). The second views AI as a system that mimics human intellect and cognitive functions, such as interpretation, inference-making, and learning. These definitions share similarities but also exhibit clear differences.

Both approaches stress that AI is intended to supplement rather than replace humans in challenging tasks (Mikalef & Gupta, 2021). They differ in their views on the extent to which AI can replicate human behavior. Some argue that AI can duplicate human abilities (Kolbjørnsrud et al., 2017; Wang et al., 2019), while others see AI as a tool that cannot completely replicate human talents (Wamba-Taguimdje et al., 2020). Additionally, definitions of AI vary, with some perceiving it as a scientific field (Schmidt et al., 2020) and others as an applied capability of machines or systems (Afiouni, 2019).

These diverse definitions highlight underlying assumptions and discrepancies regarding the nature and scope of AI. For the purposes of this essay, AI is viewed as an applied field enabling computers to recognize, understand, draw conclusions from, and learn from data to accomplish organizational and societal goals (Enholm et al., 2022).

AI involves the creation and use of computer systems capable of performing tasks traditionally requiring human intelligence. It is an interdisciplinary research area aiming to develop intelligent computers that mimic cognitive functions like learning, problem-solving, perception, reasoning, and decision-making in humans (Yassa et al., 2014).

AI systems possess capabilities in large-scale data analysis, pattern recognition, and prediction or decision-making. Machine learning, a process of training algorithms on data to identify patterns and make precise predictions or classifications, allows these systems to adapt and improve their performance over time (Premchaiswadi & Romsaiyud, 2012). Different types of AI include narrow AI, which focuses on specific tasks; general AI, capable of performing any intellectual task that a human can do; machine learning, enabling systems to learn from data; and deep learning, inspired by the neural networks of the human brain, achieving state-of-the-art performance in complex tasks (Li et al., 2023).

AI finds applications in various fields such as manufacturing, healthcare, finance, and transportation. It enables automation of routine tasks, improved decision-making, increased efficiency, and stimulation of creativity.

2.2 The Strategic Use of Artificial Intelligence in Businesses

Exploring AI in business prompts a revitalization of company strategy, avoiding automation of outdated models (Marr, 2019). Prioritizing an AI and data strategy is crucial, identifying key uses rather than peripheral studies for corporate success (Marr, 2019). Strategic AI implementation requires investment in data infrastructure, talent acquisition, and continuous learning (Albert, 2019). Upholding ethical AI practices like transparency, equity, and privacy protection is vital for consumer and stakeholder trust. The strategic application of AI fosters development, efficiency, and competitive advantage in today's dynamic business environment (Albert, 2019).

2.3 Artificial Intelligence in Practice

AI is transforming organizations across sectors through various real-world applications. Virtual assistants like Siri, Google Assistant, and Amazon Alexa utilize AI algorithms for natural language processing and user interaction (Palanica & Fossat, 2021). Recommendation systems personalize content in social networking, media streaming, and e-commerce (Packin, 2020). Predictive analytics uses AI to analyze historical data and provide accurate forecasts (Wood, 2020). Image and speech recognition systems have advanced significantly (Emerald Group, 2003). AI plays a crucial role in the development of autonomous vehicles (Caruso et al., 2022). In healthcare, AI aids in diagnostics and treatment planning (Wiffen, 1999). It enhances cybersecurity and fraud detection efforts (US Government Failing on Cybersecurity, 2009). Natural language processing enables computers to understand and translate human language (Sharma, 2020). Financial services leverage AI for credit scoring, algorithmic trading, fraud detection, and risk assessment (Hamdoun & Rguibi, 2019). The combination of robotics and AI enables automation and intelligent physical tasks (Loeb, 2022). Companies like Alibaba, Samsung, and Tesla are notable adopters of AI. As technology progresses, AI's impact on creativity, efficiency, and decision-making is expected to grow across industries.

Alibaba, a global business based in China, is renowned for its extensive e-commerce network, including platforms like Alibaba.com, Taobao, Tmall, and AliExpress. These platforms collectively form the world's largest e-commerce network (Marr, 2019). To provide customers with relevant products and optimize their browsing experience, Alibaba utilizes advanced AI technology (Marr, 2019). The AI technology employed by Alibaba's e-commerce platforms learns in real-time from user activities, such as purchases, product views, and website navigation, to personalize and optimize the displayed page content. The ultimate aim is to enhance the browsing experience and increase the likelihood of conversions (Marr, 2019).

Samsung, the world's leading consumer electronics company in terms of revenue, manufactures and sells an impressive 500 million connected devices annually (Marr, 2019). The company has developed its own AI personal assistant, Bixby, which offers voice-activated assistance and various functionalities across Samsung devices (Marr, 2019). Additionally, Samsung has integrated an AI-powered robotic arm into its manufacturing operations, automating and improving various aspects of the manufacturing process (Marr, 2019).

Tesla, known for its pioneering work in electric cars, has also positioned itself as a leader in autonomous vehicle technology. Through software upgrades, every Tesla vehicle has the potential to become self-driving, reflecting the company's dedication to advancing autonomous driving (Marr, 2019). Tesla is also involved in the production and sale of advanced batteries and solar panels, contributing to the sustainable energy sector (Marr, 2019). In the realm of self-driving vehicles, artificial intelligence (AI) and computer vision technology are crucial. AI algorithms process real-time road conditions, taking into account factors like the vehicle's direction, intended destination, and surrounding traffic. Computer vision enables the vehicle to analyze visual data from cameras and make informed decisions based on its environment (Marr, 2019).

2.4 *Trust*

Trust is a crucial aspect of AI technology in establishing effective partnerships between users and AI systems (Moorman et al., 1993; Panayides & Lun, 2009). It plays a pivotal role in the integration and acceptance of AI within organizations (Glikson & Woolley, 2020). Trust is defined as the willingness to rely on an exchange partner, ensuring reliable and dependable actions (Currall & Inkpen, 2002). In the context of AI, trust enables users to depend on its capabilities and facilitates decision-making (Fan et al., 2008). Factors influencing trust in AI include accuracy, transparency, ethical behavior, user experience, data usage transparency, human monitoring, error handling, and fail-safes (Chanda & Banerjee, 2022; Ferrario & Loi, 2022; Latifian, 2022; Pakurár et al., 2019; Santoni de Sio & van den Hoven, 2018; Sivill, 2019; Wu & Lü, 2009). Organizations should prioritize openness, moral behavior, user-centric design, and engagement to foster trust in AI (Marr, 2019).

2.5 Usefulness

Perceived usefulness, explored in the Technology Acceptance Model 3 (TAM3) and a Research Agenda on Interventions article (Venkatesh & Bala, 2008), refers to individuals' evaluation of the benefits they expect to derive from using a technology. It encompasses their beliefs regarding the potential impact of a new technology on their business practices and personal performance (Ajzen, 1991; Eriksson et al., 2005; Jahangir & Begum, 2008; Rahman et al., 2017). Customers' perceptions of a new technology's usefulness significantly influence their willingness to adopt it (Wilson et al., 2021). Companies must prioritize the perceived usefulness of their innovations, ensuring that customers perceive added value and improved performance compared to existing solutions (Wilson et al., 2021). Effective communication of benefits and advantages, along with transparent messaging about the unique value proposition, helps shape positive perceptions of usefulness among customers (Wilson et al., 2021). By addressing customer concerns and demonstrating the practical benefits, companies can increase willingness to embrace innovation, driving adoption and market acceptance (Wilson et al., 2021). Perceived usefulness affects technology evaluation, adoption, and usage, making it a focal point of research and interventions (Davis, 1989a, b). Factors influencing the perceived usefulness of AI include efficiency, personalization, decision support, automation, problem-solving, user experience, and satisfaction (Linder, 2012; Mican et al., 2020; Mühlemann et al., 2018; Nalis et al., 2022; Notess, 2001; Shen, 2012). Understanding user requirements and expectations is crucial in developing AI systems that optimize perceived usefulness and address specific user needs (Davis, 1989a, b).

2.6 Business Performance

Profitability is a critical factor that directly impacts a company's liquidity and cash flow, influencing its overall sustainability in the business landscape (Hughes et al., 2022). Generating profits is essential for securing financing, attracting investors, funding operations, and investing in future growth (Hughes et al., 2022). It serves as a key indicator of a business's financial health and long-term viability. Evaluating performance involves measuring a variety of factors, such as operational profit ratio, return on investment, sales growth rates, and total assets (Kemp et al., 2003). These measures provide insights into the overall health and performance of a firm in terms of productivity, financial growth, asset management, and profitability. On the other hand, knowledge management and successful technology adoption affect companies' competitive advantages and their business performance positively (Bolatan et al., 2022). Business performance assessment involves analyzing various indicators and metrics, including operational indicators and financial metrics (Amoah & Minishi-Majanja, 2022). By monitoring these performance indicators, businesses can gain valuable insights, identify areas for improvement, and make informed decisions to enhance their overall performance and achieve sustainable growth (Noruzi & Vargas-Hernández, 2010). Evaluating business performance at different levels provides a comprehensive view of the company's competitiveness, profitability, and long-term viability, benefiting stakeholders such as management, investors, employees, and customers (Noruzi & Vargas-Hernández, 2010).

3 Hypothesis Development

3.1 Impact of Trust on Business Performance

The increasing presence of machine-human interaction in the workplace poses numerous questions and challenges for employees. This evolving landscape, coupled with the growing prevalence of AI systems, is expected to reshape work dynamics and impact how performance is assessed and supervised (Richards et al., 2019). A sample study conducted in the health sector by Asan et al. (2020) demonstrated the positive impact of trust on performance. The research findings provided evidence of how having a high level of trust can significantly enhance performance outcomes. Another example paper by Ferrario and Viganò proposed an incremental trust model that encompasses both cognitive and non-cognitive trust aspects. The model is applicable to interactions between humans, as well as between humans and artificial intelligence. The study highlights that the impact of these trust factors on performance can vary across different businesses and organizations.

Research suggests that trust plays a crucial role in shaping the success of businesses. A study by Mayer et al. (1995) titled "An Integrative Model of Organizational Trust" explores the impact of trust on various aspects of organizational performance. This study is widely cited in the literature on trust in organizational settings.

Dirks and Ferrin (2002) conducted a meta-analysis of 87 studies to examine the relationship between trust and organizational performance. The authors found a positive and significant correlation between trust and various performance indicators, such as job satisfaction, employee commitment, cooperation, information sharing, and financial performance. The study highlights the importance of trust in fostering positive organizational outcomes and suggests that trust is a critical factor for achieving high levels of performance in businesses.

Based on these explanations, the author would like to posit the second hypothesis (H1) as follows:

H1: Trust in artificial intelligence has a positive effect on business performance.

3.2 *Impact of Usefulness on Trust*

Research by Amin et al. (2014) found that perceived usefulness positively impacts trust. When employees believe a new system or product provides added value, customers are more likely to trust it. Lee and Jun (2007) support this relationship, showing that perceiving a system as useful contributes to building trust. Horst et al. (2007) emphasize trust as a primary factor influencing perceived usefulness. Chinomona (2013) confirms the positive relationship between perceived usefulness and trust. Trust acts as a determining factor in the perceived usefulness of a technology. Perceived utility of artificial intelligence (AI) influences people's confidence in it. Users trust AI when they believe it meets their demands and delivers worthwhile results. Effects include perceived reliability, enhanced user experience, ethical considerations, confidence in decision-making, and adoption and acceptance. Li et al. (2015) and Wang et al. (2013) demonstrate the impact of perceived usefulness on trust in e-commerce.

Based on these explanations, the author would like to posit the second hypothesis (H2) as follows:

H2: The usefulness of artificial intelligence has a positive effect on trust in artificial intelligence.

3.3 *Impact of Usefulness on Business Performance*

The impact of received usefulness on business performance is significant (Tian, 2009). When customers perceive a product or service as useful, it positively affects various aspects of business performance. This includes increased customer satisfaction and loyalty (Nurcahyo, 2016), market differentiation and competitive advantage (Dumovic & Knowles, 2008), increased sales and revenue (Pollono & Pupkevičs, 2021), positive customer experience (Al-Mashhadani et al., 2020), improved brand reputation and trust (Qalati, 2019), and reduced customer complaints and returns (Hajihashemi, 2023).

To maximize the influence of received usefulness on corporate success, companies should focus on comprehending consumer wants, conducting market research, obtaining client feedback, and continually enhancing their offers. This can boost customer satisfaction, loyalty, and overall performance. Jiang et al. (2002) conducted a meta-analysis and found a positive correlation between information technology (IT) usefulness and business performance. Their research indicates that when IT systems are perceived as useful by employees, they can have a positive impact on overall organizational performance. Hu et al. (2006) explored the influence of perceived usefulness on business performance within the context of enterprise systems. Their empirical research revealed a positive and significant relationship between perceived usefulness, employee performance, and business performance. The findings suggest that when employees perceive enterprise systems as useful, it enhances their job performance, leading to improved overall business performance.

Based on these explanations, the author posits the following hypothesis (H3):

H3: The usefulness of artificial intelligence has a positive effect on business performance.

4 Research Methods

4.1 Survey Instrument

In accordance with existing literature, we crafted a survey to investigate the connections among trust in AI, the utility of AI, and the business performance associated with AI implementation. The questionnaire employed a five-point Likert-type scale, where respondents indicated their level of agreement on a scale from 1 (strongly disagree) to 5 (strongly agree). Multiple validation tests were conducted to ensure the questionnaire's reliability. The survey was evaluated by senior managers and engineers across diverse business domains, encompassing both active and inactive employment of artificial intelligence within their respective organizations.

4.2 Sample and Data Collection

The research employed a questionnaire as the data collection tool. The web-based questionnaire comprised two sections. The first section consisted of business-related questions, while the second section encompassed inquiries regarding employees' trust in artificial intelligence, the usefulness of artificial intelligence, and the measurement of business performance in organizations utilizing artificial intelligence. In the process of sample selection, careful consideration is given to including only middle and senior executives, and engineers who actively utilize artificial intelligence in their professional capacities. Individuals who do not incorporate AI into their job responsibilities are deliberately excluded from the sample. Exhibit 1 shows the main characteristics of the sample. The sample consists of 195 firms from a wide variety of industries, including metals, machine, and equipment (14.9%), food and tobacco (9.2%), construction, wood, furniture, and paper (8.7%), textile and apparel (14.9%), automotive and electronics (28.2%), chemicals and pharmaceuticals

(3.6%), and energy and mining (2.6%). The sample comprises large-sized enterprises (42.6%), small and medium-sized enterprises (SMEs) (57.4%). Most of the sample comprises of locally owned firms (55.4%), local and foreign ownership (17.4%) and the remainder of foreign ownership (17.4). In the sample, young firms (36.9%), middle aged firms (32.3%), and mature (30.8%) are equal in numbers (Table 1).

4.3 Measurement of Variables

The brief descriptions of the measures used in this study are provided in the ensuing subsections. Exhibit 2 shows all items of measurement variables.

Usefulness (USE): USE is the independent variable. The usefulness-related questions were derived from Venkatesh and Bala's (2008) study.

Trust (TR): TR is the first dependent variable. The trust-related questions were adapted from Gefen's (2000) work.

Business Performance (PERF): PERF is the second dependent variable. The questions measuring business performance in AI-utilizing businesses were translated and sourced from Sezen et al.'s (2002) research.

Previous utilization of these scales and questionnaires in various studies conducted by different researchers, along with the reliability and validity assessments

Table 1 Exhibit 1

Characteristics of sample firms			
Characteristics		Number	%
Industry sector	Metals industry, machinery, and equipment	29	14.9
	Food and tobacco	18	9.2
	Construction, wood, furniture, and paper	17	8.7
	Textile and apparel	29	14.9
	Automotive and electronics	55	28.2
	Chemicals and pharmaceuticals	7	3.6
	Energy and mining	5	2.6
	Other	35	17.9
Firm size (number of employees)	SMEs (less than 250)	112	57.4
	Large size (equal or more than 250)	83	42.6
Firm age (years of operation)	Young firms (less than 10 years)	72	36.9
	Middle-aged firms (10–20 years)	63	32.3
	Mature firms (More than 20 years)	60	30.8
Type of ownership	Locally owned	108	55.4
	Foreign owned	34	17.4
	Local and foreign owned	53	27.2
N		**195**	

conducted in those studies, indicate the absence of reliability issues with the scale (Table 2).

4.4 Analysis

To evaluate the postulated structural linkages in the model, the author used a structural equation modeling (SEM) approach (Exhibit 3). We used IBM SPSS AMOS for model analysis and data processing to assess our research model (Exhibit 3). The statistical program AMOS, which stands for "analysis of moment structures," is a supplement to SPSS. By extending conventional multivariate analysis methods like factor analysis, regression, correlation, and analysis of variance, it promotes research

Table 2 Exhibit 2

Assessment of the measurement model

Construct	Items	Model SRW[a]	AVE[b]	CR[c]
Trust in Ai (TR)	***TR***		**.74**	**.87**
Even if not monitored, I'd trust Artificial Intelligence to do the job right	TR1	.75		
I trust Artificial Intelligence	TR2	.90		
I believe that Artificial Intelligence are trustworthy	TR3	.91		
Ai Usefulness (USE)	***USE***		**.83**	**.93**
Using the Artificial Intelligence improves my performance in my job	USE1	.91		
Using the Artificial Intelligence in my job increases my productivity	USE2	.92		
Using the Artificial Intelligence enhances my effectiveness in my job	USE3	.88		
I find the Artificial Intelligence to be useful in my job	USE4	.80		
Business Performance with Ai (PERF)	***PERF***		**.65**	**.86**
With the trust and usefulness of artificial intelligence, we can continuously increase the market share compared to the competitors	PERF1	.80		
With the trust and usefulness of artificial intelligence, we can reduce costs of raw materials, labor, etc.	PERF2	.79		
With the trust and benefit of artificial intelligence, our sales are constantly increasing	PERF3	.83		
We are faster than competitors in developing new products with the trust and usefulness of artificial intelligence	PERF4	.80		
With the trust and usefulness of artificial intelligence, our product and service quality tend to increase continuously	PERF5	.80		

Notes:
[a]Model standardized regression weights are significant at $p < 0.01$
[b]Average variance extracted
[c]Composite reliability

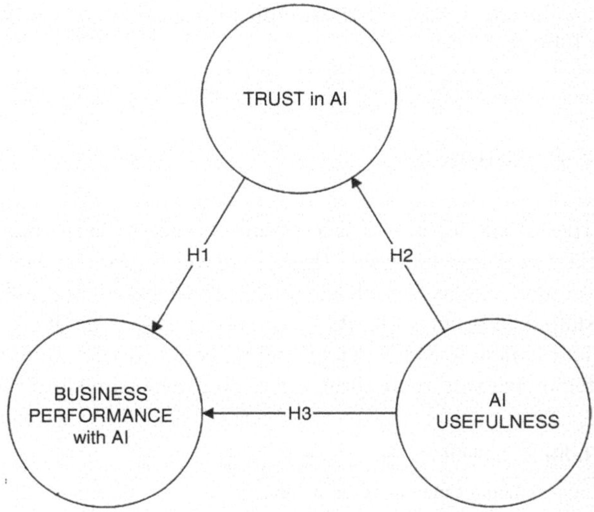

Fig. 1 The research model

and the creation of theories. A large variety of statistical and graphical capabilities are also available through AMOS (Byrne, 2016).

The reflective components were first validated using a confirmatory factor analysis (CFA). Testing for dependability was then done on each construct. We used Harman's single factor test to address any potential common method bias (CMB) that could be present. Byrne (2016)'s structural equation modeling (SEM) was used to evaluate the validity of our model. Using this method, we were able to assess how well our model matched the dataset while accounting for any moderating factors. Through route analysis and taking different mediating effects into account, the study looked at the structural links between the components. Path studies utilizing the AMOS program were used to evaluate hypotheses obtained from the conceptual model (Byrne, 2016) (Fig. 1).

4.5 Reliability and Validity of Constructs

We conducted reliability analyses using Cronbach's alpha for each of the constructs. The reliability coefficients (Cronbach's alpha) for the TR, USE, and PERF constructs were computed as 0.87, 0.93, and 0.86, respectively. The assessment of the measurement model is presented in Exhibit 2. As shown in Exhibit 2, all variables had standardized regression weights greater than 0.60, and these weights were statistically significant ($p < 0.01$) (Hair, 2009). The maximum likelihood method was utilized to estimate the model parameters. Exhibit 2 also includes Cronbach's alpha (CA) and composite reliability (CR) values, which assess the internal consistency of the constructs. Both CA and CR values exceeded 0.80, indicating satisfactory levels of construct reliability (Bagozzi & Yi, 1988). Exhibit 2 further demonstrates that the confidence reliability for TR is 0.87, for USE is 0.93, and for PERF is 0.86. Exhibit 4 presents the measures of discriminant validity. The average variance extracted (AVE) values, which are close to 0.50, indicate an acceptable level of convergent validity for the constructs in our study (Fornell & Larcker, 1981). Discriminant validity was assessed using the square root of the average variance extracted, following the recommendation of Fornell and Larcker (1981). As depicted in Exhibit 4, the square root of the AVE for each construct surpasses the inter-construct correlation values, indicating a satisfactory level of discriminant validity (Table 3).

The confirmatory factor analysis (CFA) demonstrates an excellent fit with the data, as indicated by the model fit indices falling within generally accepted ranges [$\chi 2$ statistic = 102.925, $p < 0.01$; d.f. = 51, $\chi 2$/d.f. = 2.01, GFI = 0.916, AGFI = 0.876, CFI = 0.976, TLI = 0.969, IFI = 0.976, RFI = 0.941, RMSEA = 0.072] (Hair, 2009). The maximum likelihood method was employed by the author to estimate the model parameters (Fig. 2).

4.6 *Hypotheses Testing*

A structural equation modeling (SEM) analysis was conducted to assess the study's hypotheses. The results of the structural model are presented in Exhibit 5. The model fit indices displayed, including the $\chi 2$ statistic, degrees of freedom (d.f.), $\chi 2$/d.f. ratio, GFI, AGFI, CFI, TLI, IFI, RFI, and RMSEA, fall within the accepted ranges. This indicates a good fit of the model with the data [$\chi 2$ statistic = 102.925, $p < 0.01$; d.f. = 51, $\chi 2$/d.f. = 2.01, GFI = 0.916, AGFI = 0.876, CFI = 0.976, TLI = 0.969, IFI = 0.976, RFI = 0.941, RMSEA = 0.072].

Exhibit 6 presents the results indicating that the direct relationships in the model corresponding to H2 and H3 are statistically significant. The standardized regression weights for the links between USE and TR (H2) and USE and PERF (H3) are .67 and .60, respectively (Table 4).

Table 3 Exhibit 4

Discriminant validity of the measurement model[a]			
Constructs	TR	USE	PERF
TR	.86		
USE	.61	.91	
PERF	.48	.54	.80

Notes:
[a] Italicized values on the diagonal are the square root of the AVE values

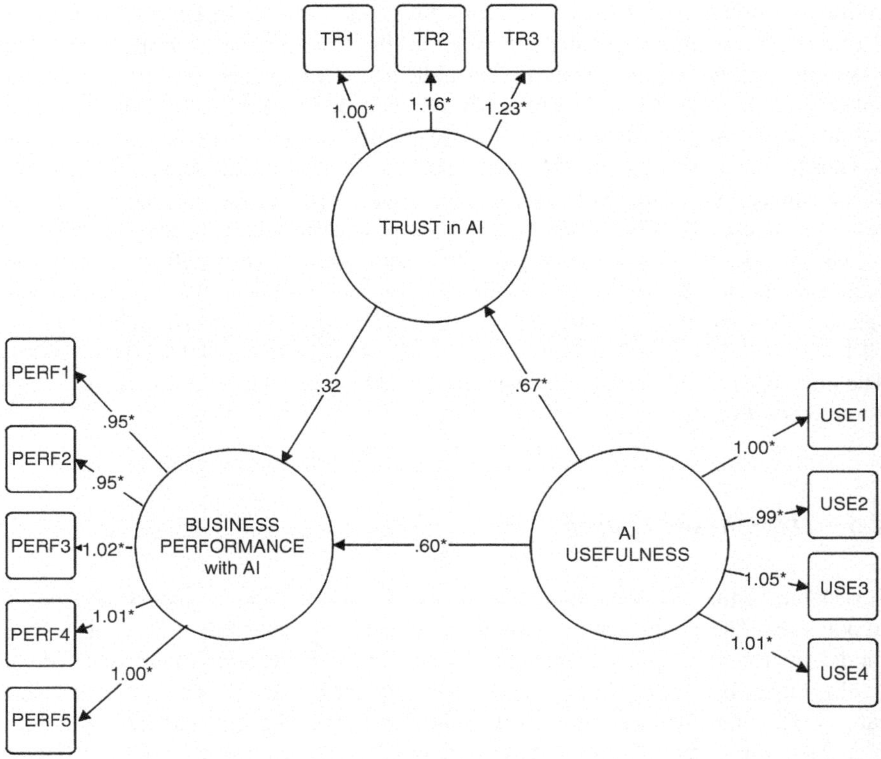

Fig. 2 Exhibit 5. Results of the mediation model *p < 0.01

5 Conclusions and Future Research

This study aims to examine the impact of artificial intelligence (AI) usage on business performance in various organizations. Through survey research and subsequent analysis using SPSS AMOS, the hypothesis suggesting a positive relationship between trust in AI and job performance was found to be unsupported. However, our findings indicate that both the usefulness of AI and trust in AI have positive effects on business performance.

The finding that the usefulness and reliability of AI positively affect business performance can be encouraging for businesses considering the adoption and implementation of this technology. Specifically, AI offers advantages such as cost savings, time savings, and improved quality by enabling more efficient decision-making in operational processes. These benefits demonstrate how businesses can enhance their competitiveness.

The ability of AI-enabled businesses to measure their own performance underscores the importance of effective and efficient utilization of this technology. For instance, AI-based forecasting models can provide significant advantages in areas

Table 4 Exhibit 6

Hypotheses testing results		
Hypotheses	Path_coefficients direct effect	Level of support
Direct effects		
H1:TR→PERF	.37	Not supported
H2:TR→USE	.67*	Supported
H3:USE→PERF	.60*	Supported

Notes:
*p < 0.00.01 (two-tailed)

like demand forecasting and inventory management. Additionally, leveraging AI technologies for crucial aspects such as customer experience and personalization can lead to increased customer satisfaction and the development of a loyal customer base.

The findings of this study serve as valuable information that can guide managers and decision-makers in adopting and effectively utilizing AI technologies within their businesses. Furthermore, understanding the role of AI technologies in business can stimulate future innovations by fostering research and development activities in this field.

In future studies, the impacts of artificial intelligence on business performance can be compared across different industries. Exploring the same topic, challenges and opportunities specific to sectors such as finance, healthcare, manufacturing, and retail can be assessed, taking into account variations in trust dynamics and performance metrics. Future research may focus on the effects of regulations on business practices, encompassing compliance costs, innovation constraints, and the role of regulatory frameworks in fostering trust. The resilience and security of artificial intelligence systems in business applications can be evaluated. On the other hand, future research endeavors may delve into potential vulnerabilities, cybersecurity challenges, and methods for ensuring the integrity and reliability of AI-supported business processes while examining the impact of artificial intelligence on business performance.

References

Afiouni, R. (2019). Organizational learning in the rise of machine learning. In *International conference on information systems, Munich, Germany.*

Ajzen, I. (1991). The theory of planned behavior. *Organizational Behavior and Human Decision Processes, 50*(2), 179–211.

Albert, E. T. (2019). AI in talent acquisition: a review of AI-applications used in recruitment and selection. *Strategic HR Review, 18*(5), 215–221

Al-Mashhadani, A. S. F., Aldori, M. K. M., & Elsayed, A. T. A. (2020, February 28). Develop a negative and positive customer experience model in online retailing by study the differences between customer experience models in successful and failure situations. *International Journal of Psychosocial Rehabilitation, 24*(04), 12–24.

Amin, M., Rezaei, S., & Abolghasemi, M. (2014). User satisfaction with mobile websites: the impact of perceived usefulness (PU), perceived ease-of-use (PEOU) and trust. *Nankai Business Review International, 5*(3), 258–274.

Amoah, G. B., & Minishi-Majanja, M. K. (2022, December 6). Key performance indicators (KPIs) in Ghanaian university libraries: developing outcome-based metrics for staff performance. *Performance Measurement and Metrics, 23*(2), 41–52.

Asan, O., Bayrak, A. E., & Choudhury, A. (2020). Artificial intelligence and human trust in healthcare: focus on clinicians. *Journal of Medical Internet Research, 22*(6), e15154.

Bagozzi, R. P., & Yi, Y. (1988). On the evaluation of structural equation models. *Journal of the Academy of Marketing Science, 16*, 74–94.

Bolatan, G. I. S., Giadedi, A., & Daim, T. U. (2022). Exploring acquiring technologies: adoption, adaptation, and knowledge management. *IEEE Transactions on Engineering Management*. https://doi.org/10.1109/TEM.2022.3168901

Byrne, B. M. (2016). *Structural equation modeling with AMOS: Basic concepts, applications, and programming*. Routledge.

Bytniewski, A., Matouk, K., Chojnacka-Komorowska, A., Hernes, M., Zawadzki, A., & Kozina, A. (2020). The functionalities of cognitive technology in management control system. In *Intelligent information and database systems: 12th Asian conference, ACIIDS 2020, Phuket, Thailand, March 23–26, 2020, Proceedings, Part II 12* (pp. 230–240). Springer.

Cammarano, A., Varriale, V., Michelino, F., & Caputo, M. (2023). A framework for investigating the adoption of key technologies: Presentation of the methodology and explorative analysis of emerging practices. *IEEE Transactions on Engineering Management, 71*, 3843–3866.

Caruso, G., Yousefi, M. K., & Mussone, L. (2022, December 15). From human to autonomous driving: A method to identify and draw up the driving behaviour of connected autonomous vehicles. *Vehicles, 4*(4), 1430–1449.

Chanda, S. S., & Banerjee, D. N. (2022, November 17). Omission and commission errors underlying AI failures. *AI & Society*. https://doi.org/10.1007/s00146-022-01585-x

Chinomona, R. (2013). The influence of perceived ease-of-use and perceived usefulness on trust and intention to use mobile social software: technology and innovation. *African Journal for Physical Health Education, Recreation and Dance, 19*(2), 258–273.

Currall, S. C., & Inkpen, A. C. (2002). A multilevel approach to trust in joint ventures. *Journal of International Business Studies, 33*, 479–495.

Davis, F. D. (1989a). Perceived usefulness, perceived ease of use, and user acceptance of information technology. *MIS Quarterly, 13*(3), 319–340.

Davis, F. D. (1989b). Technology acceptance model: TAM. In M. N. Al-Suqri & A. S. Al-Aufi (Eds.), *Information seeking behavior and technology adoption* (pp. 205–219). IGI Global.

Demlehner, Q., & Laumer, S. (2020). Shall we use it or not? Explaining the adoption of artificial intelligence for car manufacturing purposes. In *Proceedings of European conference on information systems (ECIS), Marrakech, Morocco.*

Dirks, K. T., & Ferrin, D. L. (2002). The impact of trust in organizational performance: A meta-analysis. *Academy of Management Journal, 45*(3), 446–465.

Dumovic, P., & Knowles, D. T. (2008, March). Marketing masterclass product differentiation for competitive advantage. *Journal of Medical Marketing, 8*(2), 188–188.

Emerald Group. (2003). Advancements in speech recognition technology to rejuvenate interactive voice response systems market. *Sensor Review, 23*(2).

Enholm, I. M., Papagiannidis, E., Mikalef, P., & Krogstie, J. (2022). Artificial intelligence and business value: A literature review. *Information Systems Frontiers, 24*(5), 1709–1734.

Eriksson, K., Kerem, K., & Nilsson, D. (2005). Customer acceptance of internet banking in Estonia. *International Journal of Bank Marketing, 23*(2), 200–216.

Eriksson, T., Bigi, A., & Bonera, M. (2020). Think with me, or think for me? On the future role of artificial intelligence in marketing strategy formulation. *The TQM Journal, 32*(4), 795–814.

Fan, X., Oh, S., McNeese, M., Yen, J., Cuevas, H., Strater, L., & Endsley, M. R. (2008, January). The influence of agent reliability on trust in human-agent collaboration. In *Proceedings of the 15th European conference on cognitive ergonomics: the ergonomics of cool interaction* (pp. 1–8).

Ferrario, A., & Loi, M. (2022). How explainability contributes to trust in AI. In *2022 ACM conference on fairness, accountability, and transparency (FAccT '22).*

Fornell, C., & Larcker, D. F. (1981). Structural equation models with unobservable variables and measurement error: Algebra and statistics. *Journal of Marketing Research, 18*(3), 382–388.

Gefen, D. (2000). E-commerce: the role of familiarity and trust. *Omega, 28*(6), 725–737.

Glikson, E., & Woolley, A. W. (2020). Human trust in artificial intelligence: Review of empirical research. *Academy of Management Annals, 14*(2), 627–660.

Hair, J. F. (2009). *Multivariate data analysis*. Prentice Hall.

Hajihashemi, B. (2023). Understanding and managing customer complaints. *SSRN Electronic Journal*. https://doi.org/10.2139/ssrn.4422671

Hamdoun, N., & Rguibi, K. (2019, November 20). Impact of AI and machine learning on financial industry: Application on Moroccan credit risk scoring. *Journal of Advanced Research in Dynamical and Control Systems, 11*(11-Special Issue), 1041–1048.

Horst, M., Kittschreuter, M., & Gutteling, J. M. (2007). Perceived usefulness, personal experiences, risk perception and trust as determinants of adoption of e-government services in The Netherlands. *Computers in Human Behavior, 23*(4), 1838–1852.

Hu, Q., Wei, K.-K., & Wei, J. (2006). The influence of perceived usefulness on business performance in the context of enterprise systems. *Journal of Management Information Systems, 23*(1), 115–149.

Hughes, M., Hughes, P., Hodgkinson, I., Chang, Y. Y., & Chang, C. Y. (2022). Knowledge-based theory, entrepreneurial orientation, stakeholder engagement, and firm performance. *Strategic Entrepreneurship Journal, 16*(3), 633–665.

Jahangir, N., & Begum, N. (2008). The role of perceived usefulness, perceived ease-of-use, security and privacy, and customer attitude to engender customer adaptation in the context of electronic banking. *African Journal of Business Management, 2*(1), 32–40.

Jiang, J. J., Klein, G., & Carr, C. L. (2002). The impact of IT usefulness on business performance: A meta-analysis. *Journal of Management Information Systems, 19*(2), 83–105.

Jiang, X., Jiang, X., Sun, W., & Fan, W. (2023). How do manufacturing firms manage artificial intelligence to drive iterative product innovation? *IEEE transactions on Engineering Management, 71*, 6090–6102.

Jorzik, P., Yigit, A., Kanbach, D. K., Kraus, S., & Dabić, M. (2023). Artificial intelligence-enabled business model innovation: Competencies and roles of top management. *IEEE Transactions on Engineering Management, 71*.

Kemp, R. G., Folkeringa, M., De Jong, J. P., & Wubben, E. F. (2003). *Innovation and firm performance* (Vol. No. H 200207). EIM.

Kolbjørnsrud, V., Amico, R., & Thomas, R. J. (2017). Partnering with AI: how organizations can win over skeptical managers. *Strategy & Leadership, 45*(1), 37–43.

Latifian, A. (2022, January 6). How does cloud computing help businesses to manage big data issues. *Kybernetes, 51*(6), 1917–1948.

La Torre, D., Colapinto, C., Durosini, I., & Triberti, S. (2021). Team formation for human-artificial intelligence collaboration in the workplace: a goal programming model to foster organizational change. *IEEE Transactions on Engineering Management, 5*, 1966–1976.

Lee, T., & Jun, J. (2007). Contextual perceived value?: Investigating the role of contextual marketing for customer relationship management in a mobile commerce context. *Business Process Management Journal, 13*(6), 798–814.

Li, D., Xie, L., Wang, Z., & Yang, H. (2023). Brain emotion perception inspired EEG emotion recognition with deep reinforcement learning. *IEEE Transactions on Neural Networks and Learning Systems*, 1–14.

Li, X., Liu, S., & Liang, T.-P. (2015). The impact of usefulness on trust in e-commerce: A literature review and meta-analysis. *Information & Management, 52*(3), 298–307.

Linder, M. (2012). A problem-solving perspective on strategies for appropriating environmental value some implications from considering institutional solutions to social dilemmas. *International Journal of Innovation and Sustainable Development, 6*(2), 164.

Loeb, G. E. (2022, July 5). Developing intelligent robots that grasp affordance. *Frontiers in Robotics and AI, 9*.

Makarius, E. E., Mukherjee, D., Fox, J. D., & Fox, A. K. (2020). Rising with the machines: A sociotechnical framework for bringing artificial intelligence into the organization. *Journal of Business Research, 120*, 262–273.

Marr, B. (2019). *Artificial intelligence in practice: how 50 successful companies used AI and machine learning to solve problems*. Wiley.

Mayer, R. C., Davis, J. H., & Schoorman, F. D. (1995). An integrative model of organizational trust. *Academy of Management Review, 20*(3), 709–734.

Mican, D., Sitar-Tăut, D. A., & Moisescu, O. I. (2020, December). Perceived usefulness: A silver bullet to assure user data availability for online recommendation systems. *Decision Support Systems, 139*, 113420.

Mikalef, P., & Gupta, M. (2021). Artificial intelligence capability: Conceptualization, measurement calibration, and empirical study on its impact on organizational creativity and firm performance. *Information & Management, 58*(3), 103434.

Moorman, C., Deshpande, R., & Zaltman, G. (1993). Factors affecting trust in market research relationships. *Journal of Marketing, 57*(1), 81–101.

Mühlemann, S., Kraus, R. D., Hämmerle, C. H. F., & Thoma, D. S. (2018, October). Is the use of digital technologies for the fabrication of implant-supported reconstructions more efficient and/or more effective than conventional techniques: A systematic review. *Clinical Oral Implants Research, 29*, 184–195.

Nalis, I., Kubicek, B., & Korunka, C. (2022, April 25). Resources to respond: A career construction theory perspective on demands, adaptability, and career crafting. *The Career Development Quarterly, 70*(2), 138–152.

Noruzi, M. R., & Vargas-Hernández, J. G. (2010, March 19). How intellectual capital and learning organization can foster organizational competitiveness? *International Journal of Business and Management, 5*(4).

Notess, M. (2001, August). Usability, user experience, and learner experience. *ELearn, 2001*(8), 3.

Nurcahyo, B. (2016, January 25). The role of customer satisfaction in a relation of experiential marketing and customer loyalty. *Advances in Social Sciences Research Journal, 3*(1).

Ozpamuk, M., Bolatan, G. I., VanDerSchaaf, H., & Daim, T. (2023). Exploring empowerment as an enabler of effective knowledge management. *Journal of Knowledge Management, 27*(9), 2434–2461.

Packin, N. G. (2020). Disability discrimination using AI systems, social media and digital platforms: Can we disable digital bias? *SSRN Electronic Journal*. https://doi.org/10.2139/ssrn.3724556

Pakurár, M., Benedek, S., Popp, J., Magda, R., & Oláh, J. (2019, June). Trust or doubt: Accuracy of determining factors for supply chain performance. *Polish Journal of Management Studies, 19*(1), 283–297.

Palanica, A., & Fossat, Y. (2021, May 19). Medication name comprehension of intelligent virtual assistants: A comparison of Amazon Alexa, Google Assistant, and Apple Siri between 2019 and 2021. *Frontiers in Digital Health, 3*.

Panayides, P. M., & Lun, Y. V. (2009). The impact of trust on innovativeness and supply chain performance. *International Journal of Production Economics, 122*(1), 35–46.

Pollono, E., & Pupkevičs, R. (2021, November 30). Sales variance analysis: how state-of-the-art analytical tools can contribute to increased profitability. *Journal of Revenue and Pricing Management, 22*(1), 57–71.

Premchaiswadi, W., & Romsaiyud, W. (2012, November 2). Optimizing and tuning map reduce jobs to improve the large-scale data analysis process. *International Journal of Intelligent Systems, 28*(2), 185–200.

Qalati, S. A. (2019, July). Influence of brand equity on brand trust: Mediation of brand reputation and moderation of social media. *European Journal of Business and Management*. https://ssrn.com/abstract=3693303

Rahman, M. M., Lesch, M. F., Horrey, W. J., & Strawderman, L. (2017). Assessing the utility of TAM,TPB, and UTAUT for advanced driver assistance systems. *Accident Analysis and Prevention, 108*, 361–373.

Richards, G., Yeoh, W., Chong, A. Y. L., & Popovic, A. (2019). Business intelligence effectiveness and corporate performance management: an empirical analysis. *Journal of Computer Information Systems, 59*(2), 188–196.

Santoni de Sio, F., & van den Hoven, J. (2018, February 28). Meaningful human control over autonomous systems: A philosophical account. *Frontiers in Robotics and AI, 5*.

Schmidt, R., Zimmermann, A., Möhring, M., & Keller, B. (2020). Value creation in connectionist artificial intelligence—A research agenda. In *AMCIS 2020 proceedings—Advancings in information systems research: August 10–14, 2020, Online* (pp. 1–10).

Sezen, B., Yilmaz, C., & Bağlayici, B. A. (2002). The binding role of the logistic function between marketing and production units and its effects on business performance. *Dokuz Eylül University Journal of Faculty of Economics and Administrative Sciences, 17*(2), 133–146.

Sharma, S. (2020, August 1). Natural language processing in AI: Language interpreter. *Journal of Artificial Intelligence Research & Advances, 7*(2), 21–27.

Shen, L. (2012, June 11). Perception and information behaviour of institutional repository end-users provides valuable insight for future development. evidence based library and information. *Practice, 7*(2), 81–83.

Shen, M., Gu, A., Kang, J., Tang, X., Lin, X., Zhu, L., & Niyato, D. (2023). Blockchains for artificial intelligence of things: A comprehensive survey. *IEEE Internet of Things Journal, 10*(16), 14483–14506.

Sivill, T. (2019, August 16). Ethical and statistical considerations in models of moral judgments. *Frontiers in Robotics and AI, 6*.

Tian, Y. (2009). The impact of expectations manipulation on the usefulness of analyst forecasts in firm valuation. *SSRN Electronic Journal*. https://doi.org/10.2139/ssrn.1464467

US Government Failing on Cybersecurity. (2009, November). *Computer Fraud & Security, 2009*(11), 2–3.

Venkatesh, V., & Bala, H. (2008). Technology acceptance model 3 and a research agenda on interventions. *Decision Sciences, 39*(2), 273–315.

Wamba-Taguimdje, S. L., Fosso Wamba, S., Kala Kamdjoug, J. R., & Tchatchouang Wanko, C. E. (2020). Influence of artificial intelligence (AI) on firm performance: the business value of AI-based transformation projects. *Business Process Management Journal, 26*(7), 1893–1924.

Wang, H., Huang, J., & Zhang, Z. (2019, May). The impact of deep learning on organizational agility. In *ICIS*.

Wang, Y.-S., Wang, Y.-M., Lin, H.-H., & Tang, T.-I. (2013). The role of perceived usefulness in building online trust: A comparative study of B2C and C2C contexts. *Information Systems Frontiers, 15*(2), 243–256.

Wiffen, P. (1999, March). Cancer patients overestimate their survival probabilities, which influences their preferences for treatment. *Evidence-Based Healthcare, 3*(1), 25.

Wilson, N., Keni, K., & Tan, P. H. P. (2021). The role of perceived usefulness and perceived ease-of-use toward satisfaction and trust which influence computer consumers' loyalty in China. *Gadjah Mada International Journal of Business, 23*(3), 262–294.

Wood, D. A. (2020, September). Country-wide German hourly wind power dataset mined to provide insight to predictions and forecasts with optimized data-matching machine learning. *Renewable Energy Focus, 34*, 69–90.

Wu, J. B., & Lü, G. (2009, October 9). Trust and reputation evaluation for Web services based on user experience. *Journal of Computer Applications, 29*(8), 2291–2292.

Yassa, M. M., Hassan, H. A., & Omara, F. A. (2014, March 1). Utilizing common KADS as problem-solving and decision-making for supporting dynamic virtual organization creation. *IAES International Journal of Artificial Intelligence (IJ-AI), 3*(1), 1.

Artificial Intelligence, Business Activity and Entrepreneurial Opportunities. The European Case

Francisco del Olmo-García, Fernando Javier Crecente-Romero, María Sarabia-Alegría, and María Teresa del Val Núñez

Abstract Artificial Intelligence has become one of the biggest disruptions of the last decades. Its use in the social and business spheres is increasing at a speed not seen before. These technologies mean an increase in efficiency in business processes, but also in the entrepreneurial opportunities that are perceived in the market. The aim of this paper is to study the level of implementation of Artificial Intelligence technologies in European companies and their main uses. It also seeks to explore the implications of this field for entrepreneurial opportunities and activity.

1 Introduction

The strong take-off of Artificial Intelligence (AI) in economic activity in recent years is due to a compendium of factors that have changed the economic and social reality, including the intensive use of mobile devices, the exponential increase in data generated at all times, advances in computational capacity and data storage and advances in Data Science techniques.

Thanks to this technology, AI has become known to the public, and interest in how it works and its implications has increased significantly. However, as mentioned above, AI technologies are not new to European companies. On the contrary, processes based on these technologies have been incorporated over the last few years and in a wide range of sectors. In fact, the concern about the consequences of the intensive use of AI on employment is one of the most important debates in recent years (Acemoglu et al., 2022; Acemoglu & Restrepo, 2020; Albanesi et al., 2023; Frey & Osborne, 2017; Georgieff & Hyee, 2022; Guarascio et al., 2023;

F. del Olmo-García (✉) · F. J. Crecente-Romero · M. Sarabia-Alegría · M. T. del Val Núñez
Department of Economics and Business Administration, Institute of Economic and Social Analysis (IAES), University of Alcalá, Alcalá de Henares, Madrid, Spain
e-mail: francisco.olmo@uah.es; fernando.crecente@uah.es; maria.sarabia@uah.es; mteresa.val@uah.es

M. T. Del Val Núñez et al. (eds.), *Artificial Intelligence and Business Transformation*, Contributions to Management Science,
https://doi.org/10.1007/978-3-031-58704-7_6

Valentini et al., 2023), increasing in importance as these technologies are incorporated into more economic processes and sectors.

AI has therefore become a factor of competition in geopolitical terms, with the United States and China being the main players in this rivalry (Allison & Schmidt, 2020; Hine & Floridi, 2022; Kosal, 2021).

However, the European Union, aware of the strategic potential of investing in AI technologies and the fact that it cannot be left behind in this geostrategic area, has designed a digital strategy that combines a philosophy based on technology at the service of people and companies but also contributing to achieving climate neutrality objectives by 2050.

The objectives to be achieved,[1] based on digital sovereignty, infrastructures, technology and the implementation of data regulation (especially with regard to privacy), have been established on the basis of an investment of 250 billion euros to boost digitalisation based on European NextGenerationEU funds, 43 billion euros of investments driven by policies that will support the Chips Act until 2030 and ensure that 80% of the European population has basic digital skills by 2030.

Businesses are therefore an essential player in Europe's digitisation strategy. This is because European companies are adopting AI technologies in order to improve their competitiveness (European Economic and Social Committee, 2021; Hurduzeu et al., 2022; Mikalef et al., 2023; Verbeek & Lundqvist, 2021), as digitisation allows internal and external processes to be improved and even integrated into new business models (Bouncken et al., 2021). This is aided by economic globalisation, which enhances technology transfer and its spill-over effects (Skare & Ribeiro, 2021). Nevertheless, considering this global scope, the ethical principles that should govern the adoption of AI technologies are fundamental, as pointed out by Roberts et al. (2021) and Roberts et al. (2023).

Furthermore, beyond the general orientation towards entrepreneurial activity, from an entrepreneurial activity perspective, Galindo-Martín et al. (2023) point out that entrepreneurial activity can be estimated when an environment conducive to digitisation and investment in human capital is generated that allows benefiting from digital technologies. Part of the literature points, in this sense, to the importance of coworking spaces (Bouncken et al., 2020).

Obschonka and Audretsch (2020) provide interesting conclusions in this regard. On the one hand, they point out that AI can be an external enabler of new entrepreneurial activity, considering that it could be an example of radical external changes that enhance entrepreneurial activity, allowing, in turn, the introduction of new products or services or a social focus. They also point out that limitations in relation to resources such as access to data or the cloud could generate a certain advantage in intra-entrepreneurship projects in large companies compared to independent entrepreneurial projects.

[1]A Europe fit for the digital age: https://commission.europa.eu/strategy-and-policy/priorities-2019-2024/europe-fit-digital-age_en

On the other hand, these authors highlight that AI could influence the way in which individuals participate in the entrepreneurial process. Finally, they emphasise the importance of AI in education and training within companies.

Finally, Giuggioli and Pellegrini (2023) conduct a literature review analysing the impact of AI on entrepreneurship, highlighting the impact in four areas: opportunities, decision-making, performance, and education and research.

In this sense, this paper seeks to provide an overview of business and entrepreneurial activity in the European Union from the point of view of AI technologies. To this end, firstly, the level of use of AI in European companies is shown, disaggregating the analysis from a regional point of view, but also from the point of view of company size. Secondly, a specific analysis is developed for European entrepreneurial activity. Finally, the main conclusions of the study are presented.

2 The Use of AI Technologies in European Companies

The European business sector has become a driving force to achieve the objectives defined in the European strategy. In fact, there are several sectors in which companies are leading the inclusion of AI technologies in their processes. However, there is still a long way to go if we compare the European Union with other geographies. Thus, according to Righi et al. (2022), the AI Watch Index 2021 data for the industrial perspective shows that, in the case of the European Union, 93% of AI companies do not hold patents in this field, compared to 0.8% that do and 6.2% of other types of companies that hold AI patents. In the case of the United States, 92.5% of AI companies do not hold patents in this area, compared to 1.7% that do and 5.7% of other companies that also hold patents on these technologies. More disruptive, however, is the comparison with the case of China, where only 17% of AI companies do not have patents compared to 2.3% that do. However, 80.7% of companies not considered AI companies have patents on these technologies, in contrast to the North American and European reality.

As the European Investment Bank's Digitalisation in Europe 2020–2021 survey (2021) points out, most European companies surveyed consider that the COVID-19 pandemic has accelerated the use of digital technologies, while they believe that being a leader in their adoption can be key to lasting competitiveness, but fear that Europe will be relegated to a trailing position in this area. In fact, the companies surveyed report that 37% of EU companies had not yet adopted any advanced digital technology by 2020, compared to 27% in the US, with small businesses leading the gap between these two regions. This is despite the productivity gains and skilled job creation resulting from digital transformation. To close this gap, the companies surveyed say that the diffusion of fast connections needs to be accelerated, considering that funding can be a barrier to digital transformation for small businesses.

Also, from a regional convergence perspective, it is noted that there is a strong divergence in Europe between countries that are at the forefront of the digital transformation and others that are likely to lag behind, especially among the least developed. Eurostat data on the use of AI by European companies shed light on this issue.

Thus, companies in the European Union that in 2021 used at least one of the following technologies: text mining, speech recognition, natural language generation, image recognition or image processing, machine learning for data analysis, AI-based software robotic process automation or technologies enabling machines to physically move by observing their surroundings and taking autonomous decisions, reached 6.4% for the case of companies between 10 and 49 employees (small), 12.8% in the case of companies with between 50 and 249 employees (medium) and 28.5% for the case of companies with more than 250 employees (large).[2] For the euro area, these percentages increase slightly to 7% for small companies, 14.2% for medium-sized companies and 30.8% for large companies. Therefore, the introduction of these technologies increases as the size of companies increases (taking advantage of their greater resources and capacity to hire qualified personnel), while, at the same time, relatively few companies have adapted their processes to AI, although it is noteworthy that in the case of large companies, this percentage is close to one-third of the total.

Of the companies that use AI-based technologies, the majority use commercial software or systems prepared for their use (16.9% of large companies, 6.7% of medium-sized companies and 3.1% of small companies) or use technologies developed or modified by external suppliers (15.4% of large companies, 5.2% of medium-sized companies and 2% of small companies). Also noteworthy is the percentage of companies (mainly large companies) in which, starting from commercial or open-source systems, it is the employees who modify the processes to adapt them to their needs (11.3% in the case of large companies for commercial systems and 7.7% in the case of open source systems). Finally, 10.3% of large companies choose to have their own employees develop the systems, with the percentage dropping significantly to 3.9% and 1.5% in the case of medium-sized and small companies, respectively, which, due to their size, are less able to compete for qualified talent for these functions.

From a country perspective, Fig. 1 shows how some of the Nordic countries show a higher percentage of companies that have implemented AI processes, with Denmark and Finland reaching 66.2% and 51.2% respectively in the case of larger companies. However, in the case of medium-sized companies, they are still the countries with the highest percentage, with 37.3% and 26.7% respectively. In the case of small companies, Finland does not stand out significantly, but Denmark leads the field with almost 20% of small companies using at least one of the AI technologies mentioned.

From the perspective of smaller companies, Portugal (15.7%), the Netherlands (10.1%) and Luxembourg (11.5%) also stand out. In relation to medium-sized companies, these three countries also stand out, although the cases of Slovenia (with almost 20%), Sweden (18.1%), Malta (over 17%) and Belgium (17%) are also

[2] Data is not available for companies with fewer than 10 workers.

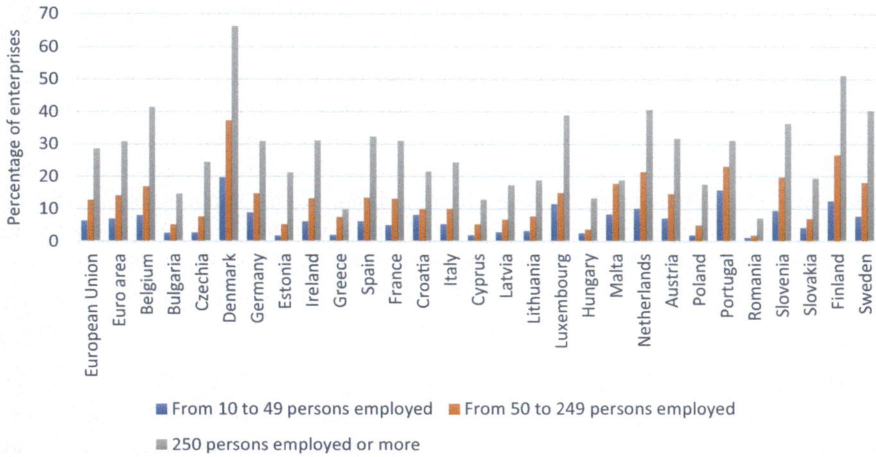

Fig. 1 % of enterprises using at least one AI technology. Source: Eurostat data

relevant. Finally, in the case of larger companies, the cases of Belgium, the Netherlands and Sweden stand out again (exceeding 40%), but also Ireland, Luxembourg, Austria, Portugal, Slovenia, Spain and Germany, which all exceed 30%.

The importance of the Nordic countries in the implementation of AI technologies in business activity has been highlighted in previous studies such as Seehuus et al. (2022), which is due to the approach these countries have adopted, claiming that responsible and ethical AI brings a competitive advantage to their productive system (Dexe & Franke, 2020; Robinson, 2020).

In this regard, the lack of global leadership in this area in Europe's largest economies stands out. Thus, the percentage of small companies that use (at least) one AI technology in their processes reaches 8.9%, while in France it is 5%, in Italy 5.1% and in Spain 6.1%. If we talk about the medium-sized companies segment, these percentages reach 14.8% for Germany, 13.1% for France and 10% and 13.4% for Italy and Spain, respectively. Finally, if we talk about large companies, Spain leads this group of larger economies with 32.3% of companies that have implemented at least one AI technology, compared to 30.9% in Germany, 31% in France and 24.3% in Italy.

Therefore, from the data observed, it can be concluded that the larger size of an economy does not necessarily imply a greater potential for its companies to adapt to the digital environment and apply AI technologies in their processes, considering, moreover, that this gap is intensified in the case of smaller companies (which are the majority in the European business structure). In fact, Spain, France and Italy do not reach the EU average for small companies, while for medium-sized and large companies, this is the case for Italy (France slightly exceeds the average).

On the other hand, beyond the general use of AI technologies, it is necessary to know what types of technologies are used by European companies. In this sense,

Fig. 2 allows us to understand the majority of technologies that have been implemented at the corporate level.

As seen in the previous analysis, it is the large companies that lead the way in the use of AI technologies in their processes. In particular, a higher percentage of both large and medium-sized companies use AI-based software robotic process automation technology (14.6% and 5.1% respectively), followed by machine learning for data analysis (13.1% and 4.4% respectively). These results make sense insofar as the first technology helps automate internal processes and improve productivity, while data analysis with machine learning algorithms helps streamline processes and achieve greater customer insight. In the case of small companies, the percentage is reversed, but by a minimal difference, although both technologies continue to lead the digitisation of companies.

Image recognition and image processing and text mining technologies also stand out for their importance in both medium-sized and large companies. In small companies, although image recognition and image processing technologies are also important, speech recognition is more important, although only in the case of text mining technologies do more than 2% of companies implement them in their processes.

Finally, it is not only important to understand the technologies used by companies, but also their usefulness in business processes. To this end, Fig. 3 provides an understanding, for the available data for 2021, of the types of activities in which the AI-related technologies analysed above are applied.

As can be seen, the main purpose for which European companies use AI technologies is related to security, both in the case of medium-sized companies (3.7% of the total) and large companies (11.2%). In the case of larger companies, production processes (9.3%), organisation of management processes (8.3%) and marketing and sales (7.2%) are also important, while in the case of medium-sized companies, the second most important purpose is related to marketing and sales processes together

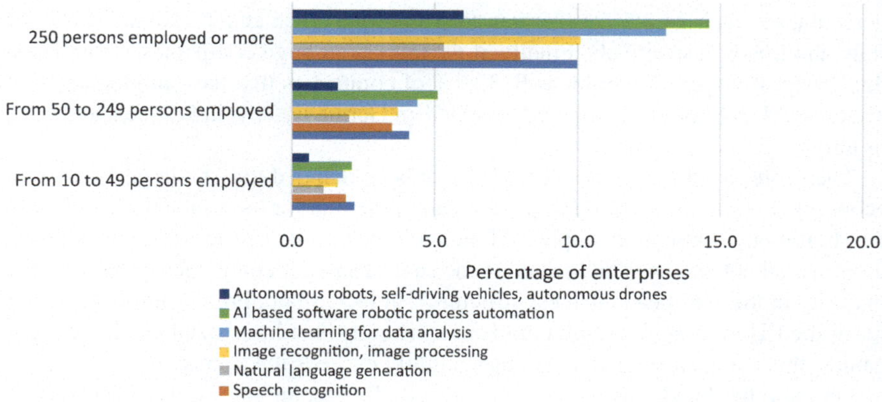

Fig. 2 Types of AI technology used. Source: Eurostat data

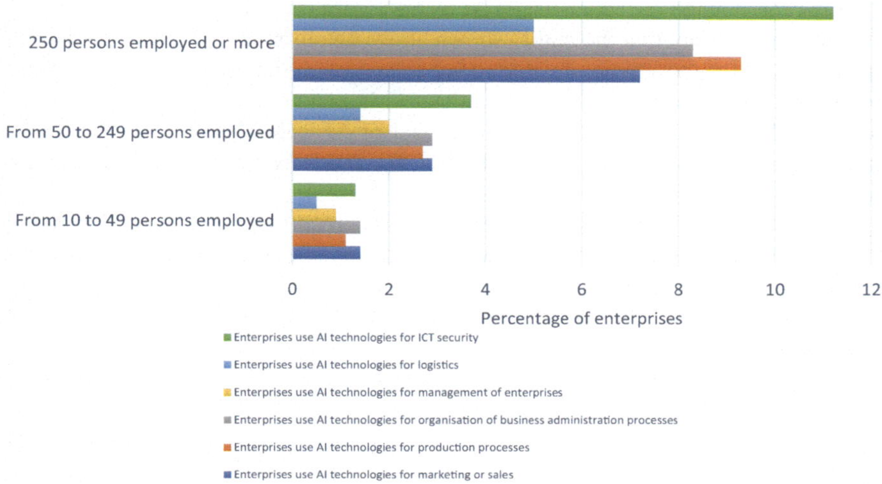

Fig. 3 Use of AI technologies. Source: Eurostat data

with the organisation of management processes (2.9% for both activities) and production processes (2.7%).

However, in the case of smaller companies, it is the processes related to management and marketing and sales that receive the most support from AI technologies (1.4%), followed by security (1.3%) and production processes (1.1%).

In short, European companies show an investment in AI-based technologies to improve their practices related to security, customer relations (marketing and sales) and to improve the productivity of their production and management processes as shown by previous literature (Kaur et al., 2023; Mihai et al., 2023; Verma et al., 2021; Vlačić et al., 2021). This behaviour is once again more prevalent among larger companies, which by nature have more resources to invest in processes that are still significantly costly.

3 AI and Entrepreneurial Opportunities in Europe

As can be seen in the previous section, the adoption of AI-based technologies is not a novelty in European companies, mainly in terms of AI-based software, robotic process automation and machine learning for data analysis, which makes sense when we understand that companies invest in this type of technology mainly to strengthen their security processes and improve the efficiency of marketing and sales, production and management processes. However, it has also been observed that the main push for these technologies is being made mainly by large companies, which tend to have greater resources in terms of investment and hiring qualified personnel.

However, beyond the actual use of AI technologies in the European business sector, there is no doubt that the reality shown above entails significant entrepreneurial opportunities for those who are able to see them, involving the generation of technology and services for all types of companies. In fact, various contributions in the academic literature reflect the positive impact of AI technologies and robotics on various economic sectors (Callarisa-Fiol et al., 2023; Wang et al., 2021) and on entrepreneurial activity (Del Olmo-García et al., 2023).

In this sense, Fig. 4 relates the entrepreneurial perception[3] collected in the Global Entrepreneurship Monitor for 2021 with companies that use at least one AI technology.[4] The case of large companies has been taken as a reference for the analysis, as they are leading the digital transformation in the European economy, as we have seen above.

The results show that entrepreneurial opportunities and business use of AI is

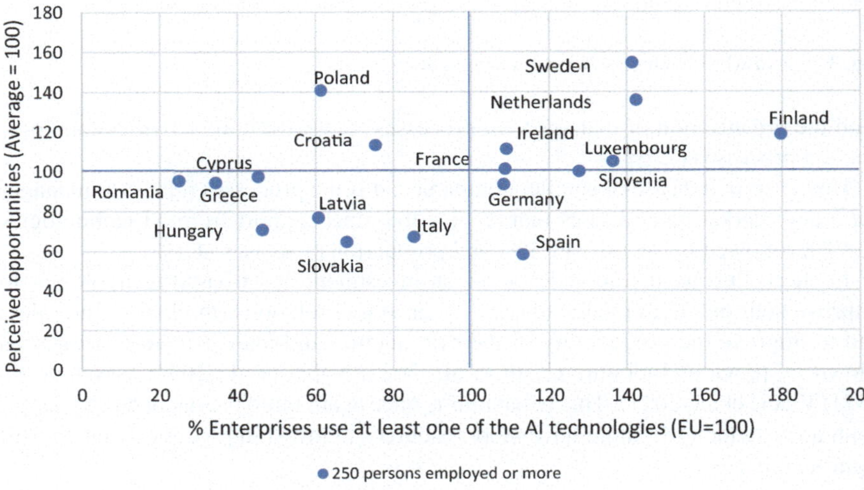

Fig. 4 Perception of entrepreneurial opportunities VS use of AI technologies. Source: Eurostat and GEM data

above average in countries such as Sweden, the Netherlands, Finland, Luxembourg and Ireland. In turn, Slovenia maintains above-average rates of AI use, but is at the

[3] Defined as the percentage of 18–64 population who see good opportunities to start a firm in the area where they live. Information in 2021 is available for Germany, Ireland, Greece, Spain, France, Croatia, Italy, Cyprus, Latvia, Luxembourg, Hungary, Netherlands, Poland, Romania, Slovenia, Slovakia, Finland, Sweden.

[4] Indices have been calculated based on 100. In the case of entrepreneurial perceptions, the average of the European countries for which information is available takes the value 100. In the case of companies that use AI technologies, the value 100 is taken as the average value of the European Union.

average in capturing entrepreneurial perceptions. However, this perception of opportunities does not translate into entrepreneurial activity per se in all cases, as shown in Fig. 5, which relates, using the same approach as the previous figure, the entrepreneurial activity[5] collected in the Global Entrepreneurship Monitor for 2021 with large companies that use at least one AI technology. In this regard, the cases of the Netherlands, Ireland and Sweden stand out.

Therefore, several conclusions can be drawn from the analysis. On the one hand, AI generates entrepreneurial opportunities, but its use is still too little implemented for these opportunities to be significant in many EU countries. On the other hand, there is still a significant gap between the capture of these opportunities and the generation of entrepreneurial activity itself, for which the importance of factors such as training, financing, and the economic and institutional framework are fundamental.

However, the possible relationship between the use of AI and general entrepreneurial activity has been analysed. However, as Obschonka and Audretsch (2020) point out, the differential capabilities of large enterprises can lead to the generation of intrapreneurial projects with competitive advantages over other entrepreneurial projects. This is why Fig. 6 shows the relationship between intrapreneurial activity[6] in the Global Entrepreneurship Monitor for 2021 and large companies that use at least one AI technology.

It can be seen that Finland, Slovenia, Ireland, Sweden and Luxembourg stand out by exceeding the average for intrapreneurial activity and for companies using at

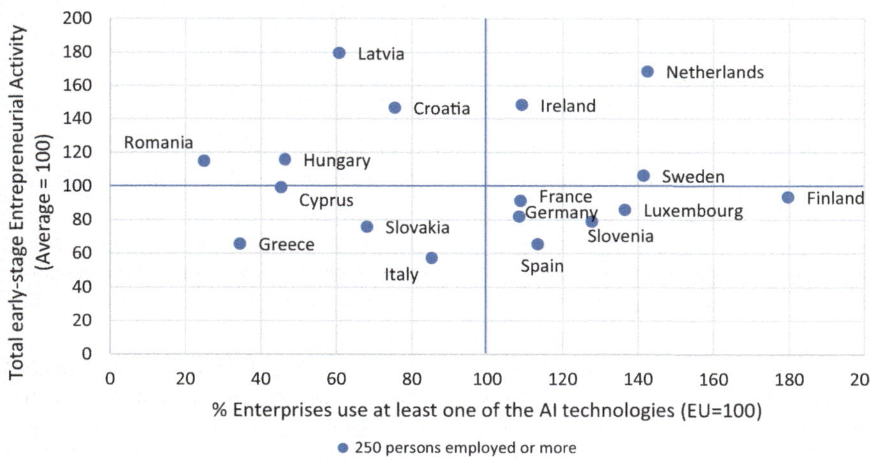

Fig. 5 Entrepreneurship VS use of AI technologies. Source: Eurostat and GEM data

[5]The Total Early-stage Entrepreneurial Activity (TEA) indicator is defined as the percentage of 18–64 population who are either a nascent entrepreneur or owner-manager of a new business.

[6]The Entrepreneurial Employee Activity indicator is defined as rate of involvement of employees in entrepreneurial activities, such as developing or launching new goods or services, or setting up a new business unit, a new establishment or subsidiary.

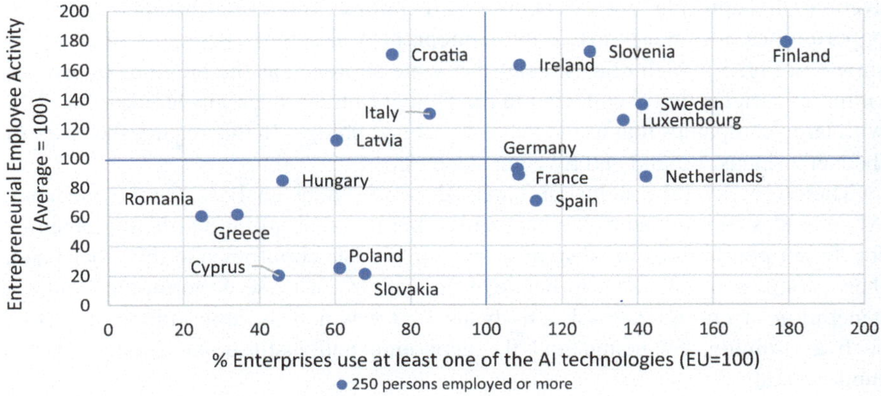

Fig. 6 Intra-entrepreneurship VS use of AI technologies. Source: Eurostat and GEM data

least one AI technology. This result is therefore more consistent with the analysis of perceptions of entrepreneurial opportunities than is the case for overall entrepreneurial activity. It seems, therefore, that the momentum that large companies are developing in terms of the use of AI generates a greater likelihood of intrapreneurial activity, since they are taking advantage of existing resources in companies and qualified personnel with the capacity to capture entrepreneurial opportunities, while at the same time solving their own needs.

4 Conclusions

There is no doubt that AI is a real disruption in the daily lives of citizens, but it is also a revolution that allows streamlining of business processes and generating new entrepreneurial opportunities.

The aim of this paper is to analyse the degree of implementation of AI technologies in the European business sector, in order to subsequently analyse the possible relationship between this degree of implementation, entrepreneurial opportunities and how these materialise in the entrepreneurial activity itself.

In this sense, European companies consider that the pandemic has accelerated the digitisation of companies, which is coupled with the implementation of the European Union's Digital Strategy. The data also show a general push by large companies in the use of AI technologies, with rates well above those of medium-sized and small companies (with countries such as Denmark and Finland standing out in particular thanks to their values and policies of considering ethical and responsible artificial intelligence as a competitive advantage, as noted by Dexe and Franke (2020) and Robinson (2020)). These results make sense considering that they have a greater capacity to invest and hire qualified personnel. Also, the main source of these technologies is obtained from commercial software or off-the-shelf systems developed or modified by external suppliers, although there are not negligible cases of companies developing or modifying products by their own employees. On the other hand, the main technologies used are AI-based software robotic process automation followed by machine learning for data analysis, which is consistent with the fact that the main applications of these technologies are to improve practices relating to safety, customer relations (marketing and sales) and to improve the productivity of production and management processes, as previous studies in the literature have highlighted (Kaur et al., 2023; Mihai et al., 2023; Verma et al., 2021; Vlačić et al., 2021).

From an entrepreneurial perspective, there is a relationship between countries with a higher percentage of large companies using AI technologies and the perception of entrepreneurial opportunities. However, these perceived opportunities are not transformed, in all cases, into entrepreneurial activity, although there is a greater relationship with respect to intra-entrepreneurial activity. This is consistent with the results shown above, as it implies a competitive advantage of the projects developed within the large companies that lead the implementation of these technologies over the rest of the entrepreneurial projects.

These results should invite public authorities to generate economic policies that encourage the implementation of AI in smaller companies (Bettoni et al., 2021; Kopka & Fornahl, 2023; Ulrich et al., 2023), while at the same time generating a framework conducive to transforming the market opportunities that these technologies offer into entrepreneurial activity (Pinchot III & Soltanifar, 2020).

References

Acemoglu, D., Autor, D., Hazell, J., & Restrepo, P. (2022). Artificial intelligence and jobs: Evidence from online vacancies. *Journal of Labor Economics, 40*(s1). https://doi.org/10.1086/718327

Acemoglu, D., & Restrepo, P. (2020). Robots and jobs: Evidence from US labor markets. *Journal of Political Economy, 128*(6). https://doi.org/10.1086/705716

Albanesi, S., Días da Silva, A., Jimeno, J. F., Lamo, A., & Wabitsch, A. (2023). *New technologies and jobs in Europe* (Working Paper Series, 31357). National Bureau of Economic Research. Available at: http://www.nber.org/papers/w31357

Allison, G., & Schmidt, E. (2020). Is China beating the U.S. to AI supremacy? Belfer Center for Science and International Affairs | Harvard Kennedy School, Paper August 2020. Available at: https://www.belfercenter.org/sites/default/files/2020-08/AISupremacy.pdf

Bettoni, A., Matteri, D., Montini, E., Gładysz, B., & Carpanzano, E. (2021). An AI adoption model for SMEs: a conceptual framework. *IFAC-PapersOnLine, 54*(1), 702–708. https://doi.org/10.1016/j.ifacol.2021.08.082

Bouncken, R. B., Kraus, S., & Martínez-Pérez, J. F. (2020). Entrepreneurship of an institutional field: the emergence of coworking spaces for digital business models. *International Entrepreneurship and Management Journal, 16*, 1465–1481. https://doi.org/10.1007/s11365-020-00689-4

Bouncken, R. B., Kraus, S., & Roig-Tierno, N. (2021). Knowledge- and innovation-based business models for future growth: digitalized business models and portfolio considerations. *Review of Managerial Science, 15*, 1–14. https://doi.org/10.1007/s11846-019-00366-z

Callarisa-Fiol, L. J., Moliner-Tena, M., Rodríguez-Artola, R., & Sánchez-García, J. (2023). Entrepreneurship innovation using social robots in tourism: a social listening study. *Review of Managerial Science, 17*, 2945–2971. https://doi.org/10.1007/s11846-023-00646-9

Del Olmo-García, F., Crecente-Romero, F. J., del Val-Núñez, M. T., & Sarabia-Alegría, & M. (2023). Entrepreneurial activity in an environment of digital transformation: an analysis of relevant factors in the euro area. *Humanities and Social Sciences Communications, 10*, 755. https://doi.org/10.1057/s41599-023-02270-0

Dexe, J., & Franke, U. (2020). Nordic lights? National AI policies for doing well by doing good. *Journal of Cyber Policy, 5*(3), 332–349. https://doi.org/10.1080/23738871.2020.1856160

European Economic and Social Committee. (2021). Boosting the use of artificial intelligence in Europe's micro, small and medium-sized enterprises. Available at: https://www.eesc.europa.eu/sites/default/files/files/qe-02-21-953-en-n.pdf

European Investment Bank. (2021). Digitalisation in Europe 2020-2021: Evidence from the EIB Investment Survey, July 2021. European Investment Bank. Available at: https://www.eib.org/attachments/efs/digitalisation_in_europe_2020_2021_en.pdf

Frey, C. B., & Osborne, M. A. (2017). The future of employment: How susceptible are jobs to computerisation? *Technological Forecasting and Social Change, 114*, 254–280. https://doi.org/10.1016/j.techfore.2016.08.019

Galindo-Martín, M. Á., Castaño-Martínez, M. S., & Méndez-Picazo, M. T. (2023). Digitalization, entrepreneurship and competitiveness: an analysis from 19 European countries. *Review of Managerial Science, 17*, 1809–1826. https://doi.org/10.1007/s11846-023-00640-1

Georgieff, A., & Hyee, R. (2022). Artificial intelligence and employment: New cross-country evidence. *Frontiers in Artificial Intelligence, 5*. https://doi.org/10.3389/frai.2022.832736

Giuggioli, G., & Pellegrini, M. M. (2023). Artificial intelligence as an enabler for entrepreneurs: a systematic literature review and an agenda for future research. *International Journal of Entrepreneurial Behavior & Research, 29*(4), 816–837. https://doi.org/10.1108/IJEBR-05-2021-0426

Guarascio, D., Reljic, J., & Stöllinger, R. (2023). *Artificial intelligence and employment: A look into the crystal ball* (GLO Discussion Paper, No. 1333). Global Labor Organization (GLO). Available at: https://www.econstor.eu/bitstream/10419/278106/1/GLO-DP-1333.pdf

Hine, E., & Floridi, L. (2022). Artificial intelligence with American values and Chinese characteristics: a comparative analysis of American and Chinese governmental AI policies. *AI & Society*. https://doi.org/10.1007/s00146-022-01499-8

Hurduzeu, G., Lupu, I., Lupu, R., & Filip, R. I. (2022). The interplay between digitalization and competitiveness: Evidence from European countries. *Societies, 157*(12). https://doi.org/10.3390/soc12060157

Kaur, R., Gabrijelčič, D., & Klobučar, T. (2023). Artificial intelligence for cybersecurity: Literature review and future research directions. *Information Fusion, 97*, 101804. https://doi.org/10.1016/j.inffus.2023.101804

Kopka, A., & Fornahl, D. (2023). Artificial intelligence and firm growth—Catch-up processes of SMEs through integrating AI into their knowledge bases. *Small Business Economics*. https://doi.org/10.1007/s11187-023-00754-6

Kosal, M. E. (2021). Artificial intelligence: Unpacking political, rhetorical, and security factors. In M. E. Kosal (Ed.), *Proliferation of weapons- and dual-use technologies. Advanced sciences and technologies for security applications*. Springer. https://doi.org/10.1007/978-3-030-73655-2_3

Mihai, F., Aleca, O. E., & Gheorghe, M. (2023). Digital transformation based on AI technologies in European Union organizations. *Electronics, 12*, 2386. https://doi.org/10.3390/electronics12112386

Mikalef, P., Islam, N., Parida, V., Singh, V., & Altwaijry, N. (2023). Artificial intelligence (AI) competencies for organizational performance: A B2B marketing capabilities perspective. *Journal of Business Research, 164*, 1–11. https://doi.org/10.1016/j.jbusres.2023.113998

Obschonka, M., & Audretsch, D. B. (2020). Artificial intelligence and big data in entrepreneurship: a new era has begun. *Small Business Economics, 55*, 529–539. https://doi.org/10.1007/s11187-019-00202-4

Pinchot III, G., & Soltanifar, M. (2020). Digital intrapreneurship. The corporate solution to a rapid digitalisation. *Digital Entrepreneurship*, 233–262. https://doi.org/10.1007/978-3-030-53914-6_12

Righi, R., Cardona, M., Samoili, S., Papazoglou, M., Vazquez-Prada, M., De Prato, G., López-Cobo, M., Dalla, A., De Nigris, S., Desruelle, P., Duch, N., Gomez, E., Hupont-Torres, I., Martinez, F., Martíne, E., Nativi, S., Nepelski, D., Pineda, C., Rossetti, F., … , Tolan, S. (2022). AI Watch Index 2021. European Commission, Joint Research Centre (JRC) [Dataset]. http://data.europa.eu/89h/e3757f41-fe54-4330-946d-ae897686164f

Roberts, H., Cowls, J., Hine, E., Mazzi, F., Tsamados, A., Taddeo, M., & Floridi, L. (2021). Achieving a 'good AI society': Comparing the aims and progress of the EU and the US. *Science and Engineering Ethics, 27*, 68. https://doi.org/10.1007/s11948-021-00340-7

Roberts, H., Cowls, J., Hine, E., Morley, J., Wang, V., Taddeo, M., & Floridi, L. (2023). Governing artificial intelligence in China and the European Union: Comparing aims and promoting ethical outcomes. *The Information Society, 39*(2), 79–97. https://doi.org/10.1080/01972243.2022.2124565

Robinson, S. C. (2020). Trust, transparency, and openness: How inclusion of cultural values shapes Nordic national public policy strategies for artificial intelligence (AI). *Technology in Society, 63*, 101421. https://doi.org/10.1016/j.techsoc.2020.101421

Seehuus, R., Täkte, C., Kuivaniemi, E., Karlsen, J. C., Linder, L., Schrode, M., & Kolari, S. (2022). The Nordic AI and data ecosystem 2022, Nordic Innovation. Available at: https://norden.diva-portal.org/smash/get/diva2:1667628/FULLTEXT02.pdf

Skare, M., & Ribeiro, D. (2021). How globalization is changing digital technology adoption: An international perspective. *Journal of Innovation & Knowledge, 6*(4), 222–233. https://doi.org/10.1016/j.jik.2021.04.001

Ulrich, P., Frank, V., & Buettner, R. (2023). Artificial intelligence in small and medium-sized family firms: An empirical study on the impact of family influence. *Corporate Governance and Organizational Behavior Review, 7*(1), 72–80. https://doi.org/10.22495/cgobrv7i1p7

Valentini, E., Compagnucci, F., Gallegati, M., & Gentili, A. (2023). Robotization, employment, and income: regional asymmetries and long-run policies in the Euro area. *Journal of Evolutionary Economics*, 1–35. https://doi.org/10.1007/s00191-023-00819-5

Verbeek, A., & Lundqvist, M. (2021). Artificial intelligence, blockchain and the future of Europe: How disruptive technologies create opportunities for a green and digital economy, The European Commission—European Investment Bank. Available at: https://www.eib.org/attachments/thematic/artificial_intelligence_blockchain_and_the_future_of_europe_report_en.pdf

Verma, S., Sharma, R., Deb, S., & Maitra, D. (2021). Artificial intelligence in marketing: Systematic review and future research direction. *International Journal of Information Management Data Insights, 1*(1), 100002. https://doi.org/10.1016/j.jjimei.2020.100002

Vlačić, B., Corbo, L., Costa, S., & Dabić, M. (2021). The evolving role of artificial intelligence in marketing: A review and research agenda. *Journal of Business Research, 128*, 187–203. https://doi.org/10.1016/j.jbusres.2021.01.055

Wang, M., Pan, C., & Ray, P. K. (2021). Technology entrepreneurship in developing countries: role of telepresence robots in healthcare. *IEEE Engineering Management Review, 49*(1), 20–26. https://doi.org/10.1109/EMR.2021.3053258

Productivity Improvements Triggered by Robotization and Internationalization Processes: The Spanish Experience

Raquel Marín, Francisco J. Santos-Arteaga, Madjid Tavana, and Debora Di Caprio

Abstract The robotization and digitalization of businesses are slow and demanding processes requiring significant investments in human capital and infrastructures. The ability of firms to adapt their production and innovation structures conditions the successful implementation of robotics and centralized computing across their value chains. At the same time, the subsequent assimilation of these technologies within their internal R&D activities should enhance firms' capacity to develop technological innovations. We study the computerization and robotization processes followed by manufacturing firms in Spain over 20 years. In particular, we analyze the effects on productivity derived from the introduction of robotics and centralized computer control systems by firms of different sizes across technological sectors of varying complexity. The interactions between these processes and firms' investment in R&D activities are also evaluated through the different considered technological and structural variables. We illustrate how, contrary to previous findings, R&D

R. Marín
Instituto Complutense de Estudios Internacionales, Universidad Complutense de Madrid, Madrid, Spain

Departamento de Análisis Económico y Economía Cuantitativa, Universidad Complutense de Madrid, Madrid, Spain
e-mail: rmarinsa@ucm.es

F. J. Santos-Arteaga
Departamento de Análisis Económico y Economía Cuantitativa, Universidad Complutense de Madrid, Madrid, Spain
e-mail: fransant@ucm.es

M. Tavana (✉)
Business Systems and Analytics Department, La Salle University, Philadelphia, PA, USA

Business Information Systems Department, Faculty of Business Administration and Economics, University of Paderborn, Paderborn, Germany
e-mail: tavana@lasalle.edu

D. Di Caprio
Department of Economics and Management, University of Trento, Trento, Italy
e-mail: debora.dicaprio@unitn.it

M. T. Del Val Núñez et al. (eds.), *Artificial Intelligence and Business Transformation*, Contributions to Management Science,
https://doi.org/10.1007/978-3-031-58704-7_7

intensity does not improve labor productivity. In the setting analyzed, productivity improvements result from robotizing production processes and exporting firms' natural evolution. A set of policy recommendations based on the empirical results obtained is proposed and discussed.

1 Introduction

Exploiting emerging technologies involves transforming the organizational structure of firms, a feature that requires modifying the products and processes inherent to the value chains of firms (Matt et al., 2015). The robotization of the production process and the recent advances in the industrial applications of artificial intelligence techniques have substantially affected the aggregate and sectorial productivity of firms, particularly before the financial crisis of 2008 (Brynjolfsson & McAfee, 2012). The decrease in productivity that emerged from the crises and consistently prevailed afterward has been highlighted and documented in the literature (Syverson, 2017). This trend has been accompanied by an increasing divergence in productivity between the firms composing the technological frontier and those lagging (Andrews et al., 2016).

Thus, understanding the main factors determining firms' productivity as the technological environment evolves constitutes an important research topic, particularly when dealing with economies such as the Spanish one, whose convergence process to the technological frontier is yet to be completed (Marín et al., 2023). We will illustrate empirically how the implementation of robotization and digitalization enhancements coupled with increments in the growth of firms and the complexity and technological content of their production processes, which lead them to compete in the international market, have a sustained positive effect on their productivity.

2 Variables

The complementarities existing in knowledge spillovers between R&D activities and the subsequent increments in the innovation and productivity of the corresponding firms have been consistently documented in the literature (Audretsch & Belitski, 2020; Audretsch et al., 2021). In particular, the introduction of robotic devices has been shown to improve the performance and productivity of Spanish manufacturing firms when comparing samples in 2008 and 2015 (Ballestar et al., 2020). A recent review of the literature describing the improvement in firm performance and efficiency that results from robotization is provided by Lakshmi and Bahli (2020).

The analysis performed in the current paper will account for and evaluate the above relationships while focusing on the influence that robotization and internationalization processes have had on the productivity of manufacturing firms in Spain over 20 years. That is, firms' productivity defines the dependent variable of the corresponding econometric model. A series of independent variables are traditionally considered when formalizing and evaluating the determinants of productivity across firms. A complete description of the independent variables selected and the literature supporting their respective inclusion follows.

2.1 *Robotization*

The robotization and introduction of digital technologies in the production processes of manufacturing firms and the subsequent effects on their efficiency and productivity have been consistently analyzed in the literature (Büyüközkan & Göçer, 2018; Liu et al., 2022; Rad et al., 2022). The main consequences regarding modifications in the industrial processes of the corresponding firms and the emergence of digital value chains within Industry 4.0 and 5.0 environments constitute a particularly relevant area of analysis (Papadopoulos et al., 2022; Vadana et al., 2020).

2.2 *R&D Intensity*

Firms perform R&D activities to obtain knowledge that can be used to foster their innovation processes and subsequent productivity (Hammar & Belarbi, 2021; Roper & Turner, 2020). R&D is a fundamental factor conditioning the increase in productivity through innovation and the capacity of firms to absorb and implement new technologies (Kim & Kim, 2022; Kose & Sakata, 2019). Furthermore, the capacity of firms to assimilate information and learn is essential for their development and expansion into the international market (Santos-Arteaga et al., 2019).

2.3 *Export Intensity*

Firms' performance and innovation capacity are generally conditioned by their interactions in global markets (Fosfuri & Tribó, 2008). The international competition exporter firms face is expected to foster their incentives to innovate and improve their productivity (Rachinger et al., 2018). However, this relationship is quite heterogeneous in practice and generally conditioned by the country and industrial sector considered (Xie & Li, 2017). In this regard, the literature on international trade encourages the development of production sectors of higher technological intensity since they display a larger contribution to the growth of total factor productivity (Zapata et al., 2023).

2.4 Technological Content of the Sector

Firms in high-tech sectors tend to invest more in R&D activities while obtaining higher productivity gains from their research (Siliverstovs, 2016). In a related paper, Radicic and Pinto (2019) illustrated the persistence of technological innovations across a set of Spanish manufacturing firms. These authors validated the finding that higher technological intensity industries foster their propensity to innovate and subsequent increments in productivity by establishing reliable cooperation networks with their suppliers (Un & Asakawa, 2015).

2.5 Size

It is generally assumed that the innovation capacity of small firms is lower than that of larger ones (Fang et al., 2021; Masood & Sonntag, 2020). Multinational enterprises are endowed with a higher capacity to finance their R&D activities (Estensoro et al., 2022; Horváth & Szabó, 2019), a feature particularly relevant when fostering innovations and productivity improvements (Müller et al., 2021). For instance, Radicic and Petković (2023) described increments in labor productivity as German Small and Medium-Sized Enterprises (SMEs) increased in size. Despite these differences, SMEs have a higher adaptation capacity than multinationals to deal with changes in the market (Radicic & Pugh, 2017).

Table 1 summarizes the signs expected to be obtained for the regression coefficients assigned to each variable.

The conceptual framework describing the relationships analyzed is summarized in Fig. 1. We have distributed the set of regressors in two categories. Those directly related to firms' investments to improve their performance define the technological framework. The structural framework is the variables conditioning the direct interactions of firms with the market environment.

The main results derived from the empirical analysis are determined by the variables defining the technological framework. The hypotheses tested can be summarized through the following relation, with the remaining variables acting as structural determinants:

H1: Robotization and investment in R&D positively affect firm productivity.

Table 1 Expected sign of the regression coefficients

	Predictors				
Dependent variable	Robotization	R&D	Exports	Tech content	Size
Productivity	+	+	+	+	+

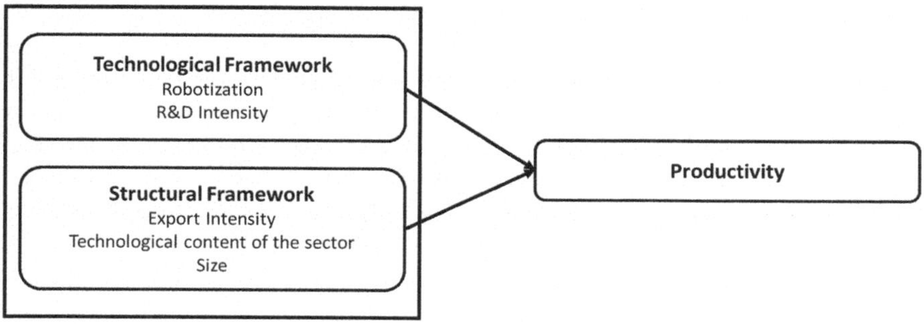

Fig. 1 Basic conceptual framework

3 Data and Methodology

A random effects econometric model describing the relationship between robotization and productivity will be carried out using data on manufacturing firms retrieved from the Survey on Business Strategies (Encuesta sobre Estrategias Empresariales—ESEE) conducted periodically since 1990 under the auspices of the Spanish Ministry of Finance and Public Function. The main features of the sample were analyzed together with a description and justification of the econometric model applied.

3.1 Sample Features

ESEE provides a large amount of micro-level information for firms with more than ten employees located in Spain and operating within 20 manufacturing industries from 1991 to 2017. In particular, the database includes all the firms with more than 200 employees and a representative sample of those employing between 10 and 200 workers. The selection of the latter is based on a random stratified sampling scheme.

ESEE contains specialized information on firms' productive and technological activities, business strategies, and performance. Since 1994, ESEE has included information regarding the implementation of robotics in firms' production processes. The survey does indeed have a specific query related to robotization. Firms are requested to answer positively or negatively to the following question: "Indicate whether your firm uses robotics in the production process." On the minus side, this information is provided only once every 4 years. Thus, the empirical analysis performed is determined by the behavior of firms through the period 1994–2014, corresponding to the data currently available.

We assess whether the robotization of productive activities has contributed to firms' productivity by analyzing a sample encompassing 99 firms that answered the corresponding survey question at any point through the period under analysis. Since the database results from a survey, firms may be unable to respond in certain years, drop out of the sample as they leave the market, or appear as new entries as they enter the market.

It must be emphasized that not all the firms studied implemented robotics within the period analyzed. We have decided to keep all the firms answering the survey question in the sample to capture the main differences between the firms introducing robotics and those that do not, as well as to control for potential overestimation effects of robotization on productivity. Specifically, 34.3% of the firms included in the sample adopted robotics in their production processes.

3.2 Econometric Model and Estimation Procedure

To identify whether and to what extent the implementation of robotics in production processes contributes to the productivity of firms—and, consequently, to their technological performance—, the following econometric model has been defined assuming that robotization is often conditioned by environmental competitive pressures and the subsequent allocation of resources to innovative activities:

$$lp_{it} = \beta_0 + \beta_1 rob_{it} + \beta_2 rdint_{it} + \beta_3 exportint_{it} + \beta_4 high_{it} + \beta_5 size_{it} + \eta_i + u_{it} \quad (1)$$

where the subscript *it* refers to firm *i* in period *t*; η_i represents firm-specific time-invariant effects and u_{it} is the error term.

The dependent variable is the firm's labor productivity (*lp*), measured as the proportion of sales by employee and expressed in logarithms to focus on the determinants of productivity growth.

The main independent variable, robotization (*rob*), captures whether or not the firm uses robotics in its production process. It takes the value 1 if the firm has indeed implemented this technological advance and 0 otherwise. The R&D intensity of firms (*rdint*), measured as the percentage of R&D investment over sales, is also included as a regressor due to the wide evidence demonstrating that the evolution of firm's productivity is conditioned by the development of technological activities. To account for environmental competitive pressures, the export intensity of firms (*exportint*) and the technological sophistication of the industry in which they operate (*high*) are also included as independent variables. The latter variable is defined according to the classification of manufacturing industries introduced by Hatzichronoglou (1997). Industries are categorized into three main groups determined by the low-, medium-, and high-technological content of the corresponding firms. We have unified the first two groups into a unique category while high technological content firms compose the complementary group, defining the subsequent binary variable.

Regarding the export intensity of firms and the heterogeneity of the findings described in Sect. 2.3, the literature highlights two main arguments that justify the greater levels of productivity displayed by exporting firms (Steinhäuser et al., 2021). The first one relates to the market selection hypothesis, according to which export markets tend to select the most efficient firms due to the greater competition faced by exporters in foreign markets. The second one considers the learning process hypothesis, according to which export markets allow firms to improve their productivity by exchanging knowledge in foreign markets. At the same time, the technological sophistication of industries may determine the dynamics of growth and competitiveness of firms since the sources of technological opportunities and the conditions required for knowledge appropriability and cumulativeness differ across sectors.

Finally, firm size (*size*) is included as a control variable to account for the main features described in Sect. 2.5 defining the behavior of competing firms of different sizes.

Table 2 provides a summary of the variables composing the model and their corresponding definitions.

The availability of panel data allows us to estimate the econometric model by applying either fixed or random effects regression techniques. The latter method accounts for unobserved heterogeneity across firms and will be used for the estimation. However, this procedure relies on the absence of correlation between the unobserved individual effects and the explanatory variables to ensure the efficiency and consistency of the estimators (Álvarez et al., 2016; Wooldridge, 2002). To choose the appropriate estimation method, the Hausman test was conducted. The results failed to reject the null hypothesis, implying that the random effects estimation procedure is the most appropriate technique.

Before discussing the estimation results, basic descriptive statistics of the sample firms categorized according to whether or not they implemented robotics within their production processes are presented in Table 3. Clearly, firms introducing robotics are more productive, export-oriented, and larger than those firms that did not introduce robotics at any point through the period analyzed. The differences between both groups of firms are more noticeable in size, with those introducing robotics being more than two times larger on average. Significant differences between both groups can also be observed regarding R&D intensity, with firms not implementing robotics displaying a more than 1.5 times greater innovative effort. This latter group also has a higher presence in more technologically intensive industries. It should, however, be emphasized that this variable exhibits the highest observed heterogeneity of the set.

All in all, firms introducing robotics in their production processes through the period considered are more productive, export-oriented, and larger but show lower innovative capabilities and have a lower presence in high-tech industries than those

Table 2 Variable definition

lp_{it}	Labor productivity (sales by employee, € thousands), firm i year t
rob_{it}	Use of robotics in production process (1 yes, 0 no), firm i year t
$rdint_{it}$	R&D expenditure (as a percentage of sales), firm i year t
$exportint_{it}$	Export volume (as a percentage of sales), firm i year t
$high_{it}$	High technological industry (1 yes, 0 no), firm i year t
$size_{it}$	Firm size (number of employees), firm i year t

Table 3 Descriptive statistics

	Firms implementing robotics		Firms not implementing robotics	
	Mean	Std. dev./mean	Mean	Std. dev./mean
lp	114.73	1.57	72.53	1.62
rdint	0.50	2.01	0.86	3.12
exportint	0.28	1.10	0.22	1.31
high	0.04	4.78	0.06	4.04
size	236.35	3.19	116.78	3.02

Table 4 Correlation matrix

Variables	1	2	3	4	5	6
1. lp	1					
2. rob	0.1826	1				
3. rdint	0.0484	–0.0669	1			
4. exportint	0.2443	0.0796	0.1566	1		
5. high	0.053	–0.0314	0.2257	0.0992	1	
6. size	0.1552	0.1074	0.1381	0.1794	0.0449	1

firms not introducing robotics. The intuition following this description will be validated—to a certain extent—by the econometric regression analysis presented in the next section.

We conclude by illustrating in Table 4 the absence of any significant correlation among the main variables whose relationships are being evaluated empirically.

4 Estimation Results

Table 5 illustrates the results obtained from the econometric regression, which are positive and significant for all the variables except the intensity of investment in R&D. Robotization, export intensity, the technological sophistication of the sector and size have a positive and significant effect on productivity. The behavior of the R&D intensity variable constitutes an exception regarding the results described in the literature. Firms increasing the intensity of their R&D investments do not receive a payoff in terms of productivity increments, a quality that is, however, derived from robotization processes. In a nutshell, firms implementing robotics and extending their domain of operation into the international market display positive productivity effects.

The profile that emerges from the empirical results presented in Table 5 describes firms whose productivity is positively affected by the robotization processes implemented, the competitive pressures arising from interacting in international markets, the technological content of the sectors within which they operate, and their relative size. More importantly, firms implementing robotics in their production processes do not improve their productivity by increasing the intensity of their investment in

Table 5 Estimation results

Independent variables	Coefficients
rob	0.4319 (0.1862)**
rdint	0.0182 (0.0244)
exportint	1.3386 (0.3383)***
high	0.6036 (0.2848)**
size	0.0004 (0.0001)***
constant	9.9180 (0.1319)***
Num. observations	195
Wald χ^2 (p-value)	53.73 (0.000)
Hausman test (p-value)	5.20 (0.274)

Robust standard errors in parenthesis
** significant at 5%; *** significant at 1%

R&D. Except for this latter quality, the resulting intuition could be directly extrapolated to the implementation of digital technologies, whose adoption is generally constrained by the capacity of firms to gather financial and capital resources (Estensoro et al., 2022). We elaborate on this latter point in the next section.

5 Policy Implications

The importance of R&D investment as a factor conditioning the increase in productivity through innovation and the capacity of firms to assimilate and implement new technologies have been challenged by the results obtained. Robotization processes combined with the natural evolution of exporting firms take on the productivity enhancement role traditionally assigned to R&D. The design of a cooperative environment and ecosystem fostering the introduction and assimilation of innovations across firms should, therefore, constitute one of the most relevant features of the Spanish technological agenda (Elia et al., 2021; Marín et al., 2023).

The technological policy debate has recently focused on the positive effect of information and communication technologies on labor productivity (Koutroumpis et al., 2020), and, more importantly, its weakening in recent years (Bruno et al., 2019). These latter authors have described the decrease in total factor and labor productivity among most European Union countries in 2008–2018 relative to 1997–2007. As follows from the analysis performed, the implementation of a successful Industry 5.0 strategy would require combining both technological and human capital, similar to how it was done when introducing robotics and technological developments along the value chain of firms (López et al., 2011; Vassilakopoulou et al., 2023).

Furthermore, the specific complementarities existing between technological and institutional environments as determinants of the evolution of countries' industrial structures and the subsequent productivity of firms must be carefully incorporated into the analysis (Yang et al., 2021). In this regard, a detailed evaluation of the relationship between technological and institutional transformations and the performance of firms in terms of innovation and productivity requires considering additional variables and alternative scenarios complementing those studied in the current paper (Santos-Arteaga et al., 2020, 2023).

6 Conclusion

We have analyzed the effect of robotization on the productivity of Spanish firms through a 20-year period. In addition to the positive effect obtained, we have also observed that larger firms facing higher competitive pressures from the export intensity and the technological sophistication of the industry within which they operate are more productive. The lack of significance displayed by firms' R&D investments constitutes a relevant result regarding the potential substitutability between robotization and internationalization processes on the one hand and the financing of internal research and development activities on the other.

A drawback of the analysis performed is that the variables accounting for the robotization process of firms are retrieved every 4 years, limiting the capacity of the sample to provide a consistent picture of the process. Similarly, the 20-year period analyzed generates substantial variability in terms of the firms composing the database, preventing a specific study of those firms developing and implementing reliable robotization strategies.

Current digitalization processes have a substantial effect on the performance of innovations in terms of productivity and their capacity to generate new knowledge. A detailed analysis of the firms undergoing these processes and a comparison with those that defined and implemented robotization strategies would constitute a potential line of research, allowing the identification of similarities and differences between both scenarios and the resulting consequences for the performance of firms.

References

Álvarez, I., Marin, R., & Santos-Arteaga, F. J. (2016). R&D internationalisation and the strategic relevance of the institutional framework in host locations. *International Journal of Management and Decision Making, 15*(3–4), 205–231.

Andrews, D., Criscuolo, C., & Gal, P. (2016). *The global productivity slowdown, technology divergence and public policy: A firm-level perspective* (OECD Background Paper). OECD

Audretsch, D. B., & Belitski, M. (2020). The role of R&D and knowledge spillovers in innovation and productivity. *European Economic Review, 123*, 103391.

Audretsch, D. B., Belitski, M., & Caiazza, R. (2021). Start-ups, innovation and knowledge spillovers. *The Journal of Technology Transfer, 46*(6), 1995–2016.

Ballestar, M. T., Díaz-Chao, Á., Sainz, J., & Torrent-Sellens, J. (2020). Knowledge, robots and productivity in SMEs: Explaining the second digital wave. *Journal of Business Research, 108*, 119–131.

Bruno, R. L., Douarin, E., Korosteleva, J. A., & Radosevic, S. (2019). *Determinants of productivity gap in the European Union: A multilevel perspective* (IZA Discussion Paper No. 12542).

Brynjolfsson, E., & McAfee, A. (2012). *Race against the machine: How the digital revolution is accelerating innovation, driving productivity, and irreversibly transforming employment and the economy*. MIT Press.

Büyüközkan, G., & Göçer, F. (2018). Digital supply chain: Literature review and a proposed framework for future research. *Computers in Industry, 97*, 157–177.

Elia, G., Margherita, A., Ciavolino, E., & Moustaghfir, K. (2021). Digital society incubator: Combining exponential technology and human potential to build resilient entrepreneurial ecosystems. *Administrative Sciences, 11*(3), 96.

Estensoro, M., Larrea, M., Müller, J. M., & Sisti, E. (2022). A resource-based view on SMEs regarding the transition to more sophisticated stages of Industry 4.0. *European Management Journal, 40*(5), 778–792.

Fang, X., Paez, N. R., & Zeng, B. (2021). The nonlinear effects of firm size on innovation: An empirical investigation. *Economics of Innovation and New Technology, 30*(1), 48–65.

Fosfuri, A., & Tribó, J. A. (2008). Exploring the antecedents of potential absorptive capacity and its impact on innovation performance. *Omega, 36*(2), 173–187.

Hammar, N., & Belarbi, Y. (2021). R&D, innovation and productivity relationships: Evidence from threshold panel model. *International Journal of Innovation Studies, 5*(3), 113–126.

Hatzichronoglou, T. (1997). *Revision of the high-technology sector and product classification* (OECD Science, Technology and Industry Working Papers, No. 1997/02). OECD. https://doi.org/10.1787/134337307632

Horváth, D., & Szabó, R. Z. (2019). Driving forces and barriers of Industry 4.0: Do multinational and small and medium-sized companies have equal opportunities? *Technological Forecasting and Social Change, 146*, 119–132.

Kim, J., & Kim, W. (2022). The intensity and diversity of R&D partner types and product development: Do product innovation types and industry sectors matter? *Technological Forecasting and Social Change, 184*, 121958.

Kose, T., & Sakata, I. (2019). Identifying technology convergence in the field of robotics research. *Technological Forecasting and Social Change, 146*, 751–766.

Koutroumpis, P., Leiponen, A., & Thomas, L. D. (2020). Small is big in ICT: The impact of R&D on productivity. *Telecommunications Policy, 44*(1), 101833.

Lakshmi, V., & Bahli, B. (2020). Understanding the robotization landscape transformation: A centering resonance analysis. *Journal of Innovation and Knowledge, 5*(1), 59–67.

Liu, Z., Liu, Q., Xu, W., Wang, L., & Zhou, Z. (2022). Robot learning towards smart robotic manufacturing: A review. *Robotics and Computer-Integrated Manufacturing, 77*, 102360.

López, S., Molero, J., & Santos-Arteaga, F. J. (2011). Poverty traps in a frictionless world: The effects of learning and technology assimilation. *Structural Change and Economic Dynamics, 22*, 106–115.

Marín, R., Santos-Arteaga, F. J., Tavana, M., & Di Caprio, D. (2023). Value Chain digitalization and technological development as innovation catalysts in small and medium-sized enterprises. *Journal of Innovation and Knowledge, 8*(4), 100454.

Masood, T., & Sonntag, P. (2020). Industry 4.0: Adoption challenges and benefits for SMEs. *Computers in Industry, 121*, 103261.

Matt, C., Hess, T., & Benlian, A. (2015). Digital transformation strategies. *Business and Information Systems Engineering, 57*, 339–343.

Müller, J. M., Buliga, O., & Voigt, K. I. (2021). The role of absorptive capacity and innovation strategy in the design of Industry 4.0 business models—A comparison between SMEs and large enterprises. *European Management Journal, 39*(3), 333–343.

Papadopoulos, T., Singh, S. P., Spanaki, K., Gunasekaran, A., & Dubey, R. (2022). Towards the next generation of manufacturing: Implications of big data and digitalization in the context of Industry 4.0. *Production Planning & Control, 33*(2–3), 101–104.

Rachinger, M., Rauter, R., Müller, C., Vorraber, W., & Schirgi, E. (2018). Digitalization and its influence on business model innovation. *Journal of Manufacturing Technology Management, 30*(8), 1143–1160.

Rad, F. F., Oghazi, P., Palmié, M., Chirumalla, K., Pashkevich, N., Patel, P. C., & Sattari, S. (2022). Industry 4.0 and supply chain performance: A systematic literature review of the benefits, challenges, and critical success factors of 11 core technologies. *Industrial Marketing Management, 105*, 268–293.

Radicic, D., & Petković, S. (2023). Impact of digitalization on technological innovations in small and medium-sized enterprises (SMEs). *Technological Forecasting and Social Change, 191*, 122474.

Radicic, D., & Pinto, J. (2019). Collaboration with external organizations and technological innovations: Evidence from Spanish manufacturing firms. *Sustainability, 11*(9), 2479.

Radicic, D., & Pugh, G. (2017). Performance effects of external search strategies in European small and medium-sized enterprises. *Journal of Small Business Management, 55*, 76–114.

Roper, S., & Turner, J. (2020). R&D and innovation after COVID-19: What can we expect? A review of prior research and data trends after the great financial crisis. *International Small Business Journal, 38*(6), 504–514.

Santos-Arteaga, F. J., Di Caprio, D., & Tavana, M. (2023). Information and communication technologies and labor productivity: A dynamic slacks-based data envelopment analysis. *Journal of the Knowledge Economy*. https://doi.org/10.1007/s13132-023-01634-w

Santos-Arteaga, F. J., Tavana, M., Torrecillas, C., & Di Caprio, D. (2020). Innovation dynamics and financial stability: A European Union perspective. *Technological and Economic Development of Economy, 26*(6), 1366–1398.

Santos-Arteaga, F. J., Torrecillas, C., & Tavana, M. (2019). Dynamic effects of learning on the innovative outputs and productivity in Spanish multinational enterprises. *The Journal of Technology Transfer, 44*, 97–131.

Siliverstovs, B. (2016). R&D and non-linear productivity growth. *Research Policy, 45*(3), 634–646.

Steinhäuser, V. P. S., Paula, F. D. O., & de Macedo-Soares, T. D. L. V. A. (2021). Internationalization of SMEs: A systematic review of 20 years of research. *Journal of International Entrepreneurship, 19*(2), 164–195.

Syverson, C. (2017). Challenges to mismeasurement explanations for the US productivity slowdown. *The Journal of Economic Perspectives, 31*(2), 165–186.

Un, C. A., & Asakawa, K. (2015). Types of R&D collaborations and product innovation: The benefit of collaborating upstream in the knowledge chain. *Journal of Product Innovation Management, 32*, 138–153.

Vadana, I. I., Torkkeli, L., Kuivalainen, O., & Saarenketo, S. (2020). Digitalization of companies in international entrepreneurship and marketing. *International Marketing Review, 37*(3), 471–492.

Vassilakopoulou, P., Haug, A., Salvesen, L. M., & Pappas, I. O. (2023). Developing Human/AI interactions for chat-based-customer-services: Lessons learned from the Norwegian government. *European Journal of Information Systems, 32*(1), 10–22.

Wooldridge, J. M. (2002). *Econometric analysis of cross-section and panel data*. MIT Press.

Xie, Z., & Li, J. (2017). Export intensity, domestic competition, and product innovation in an emerging economy. *International Journal of Technology Management, 74*(1-4), 96–121.

Yang, M., Fu, M., & Zhang, Z. (2021). The adoption of digital technologies in supply chains: Drivers, process and impact. *Technological Forecasting and Social Change, 169*, 120795.

Zapata, A. N., Arrazola, M., & de Hevia, J. (2023). Technological intensity in manufacturing trade between ASEAN and the EU: Challenges and opportunities. *Asia Europe Journal, 21*(1), 23–42.

Part II
Human Resources

Artificial Intelligence in Operations Management. A Strategy to Make Organizations More Attractive

Raquel Sastre, Alba Yela Aránega, Raúl Castaño Urueña, and Rafael Castaño Sánchez

Abstract Artificial Intelligence (AI) is gaining ground in many areas. This chapter addresses the implementation of AI within operations management. This study goes beyond the capabilities and scope of AI applications, with the aim of determining the known AI methods and tools, investigating how their strategies and effects are assessed, and exploring which ones are applied in day-to-day Operations Management (OM) in organizations. To this end, a questionnaire is developed for one hundred professionals in the logistics sector to determine the best-known AI methods and tools, those that are applied on a daily basis in organizations, and examines the ways of assessing the strategies used and the effects of their application in OM.

1 Introduction

The origin of Artificial Intelligence (AI) goes back almost a century. In 1936, Alan Turing was already theorizing about machines that function autonomously. The term Artificial Intelligence was coined in the 1950s, but a substantial shift took place at the beginning of the twenty-first century.

This century has seen the emergence of countless AI methods and tools that mimic the creative activity of humankind, not as just another activity but as the central characteristic of its reason, since man is creative by nature (Barrena, 2007). This creativity transcends the artistic field: it is related to what it means to make our

R. Sastre (✉)
Instituto Universitario ESEADE, Buenos Aires University (UBA), Faculty of Economical Sciences, Buenos Aires, Argentina

A. Yela Aránega · R. Castaño Urueña · R. Castaño Sánchez
Department of Economics and Business, Faculty of Economics, Business and Tourism, University of Alcalá, Alcalá de Henares, Madrid, Spain
e-mail: alba.yela@uah.es; raul.castano@edu.uah.es; rafael.castano@uah.es

M. T. Del Val Núñez et al. (eds.), *Artificial Intelligence and Business Transformation*, Contributions to Management Science,
https://doi.org/10.1007/978-3-031-58704-7_8

lives easier, devising solutions that help us to explore the world around us. With these ends in mind, AI emerges as combinations of algorithms created to act as the human brain would: technologies capable of automating tasks, solving problems, and "thinking" as a human being would.

Thus, advances in AI are not mere innovations: in a dialogue between the old and the new, useful methods and tools emerge that are valuable for practical life. Operations Management (OM)—understood as the practice of Planning, Executing, and Evaluating (Monitoring) the internal processes in an organization—aims to improve its performance, efficiency, and productivity. Belhadi et al. (2021), discussing OM, state that organizations are currently moving towards changing their mindset to incorporate AI techniques into their operations. Numerous success stories can be found in different areas of operations (healthcare, stock management, transportation, and logistics, among others) (Belhadi et al., 2021; Fosso Wamba et al., 2022; Yang et al., 2022).

Two centuries ago, the leap in human creativity was unimaginable in the light of everything that had gone before, and although technological advances were conceived, they were not equated with the human mind. Machines can perform the same functions as man more perfectly, but they cannot innovate, said Charles S. Peirce, the nineteenth-century philosopher. Because the machine has no imagination and no capacity for surprise, it can only simulate those capacities by applying deductions (CP 5579, 1898).

The arrival of the Internet in the home has revolutionized human behavior (Rifkin, 2014), from the way we work (teleworking) to online shopping (e-commerce) and the way we carry out tasks at home (home automation). The many applications that facilitate people's daily lives have made the Internet an indispensable tool in companies and homes (Dankan Gowda et al., 2020). Machine-to-machine interaction, also known as M2M, has begun to take place. In other words, any object can be connected to the Internet and interact without human intervention. For example, "*Smart Cities*" or "Smart Buildings" use IoT (Internet of Things) devices to improve the control of supplies such as electricity or water in a building, public transport cards, air traffic control, and road control.

This chapter goes beyond the capabilities and scope of AI applications, aiming to determine the known AI methods and tools, investigate how their strategies and effects are valued, and to explore which are applied in day-to-day Operations Management (OM) in organizations. We will answer such questions as: What tools are currently used in operations management? How can we innovate in the working environment? Is AI useful in operations departments? Is operations management advanced in the implementation of AI? Therefore, this chapter contributes to the literature review, and offers a case study of research into, and the impact of, AI within OM. It may be of interest to any company looking to incorporate AI tools into their work environment for the first time, to researchers looking to deepen their studies within the field of business operations, to anyone who is considering implementing new tools in their task processing, and to managers aiming to increase the productivity and performance of their team members.

Identifying strategies and effects involves addressing the aesthetics and ethics of AI applications, which is the subject of the first part of this chapter. This is followed by a brief description of the predominant AI methodologies that can be applied to OM in organizations. The results are then presented, followed by conclusions. There is still much to be developed in Artificial Intelligence in Operations Management: we cannot know what future tools might facilitate the improvement of the productivity of organizational resources. In the meantime, the application of AI in OM favors the growth of organizations because it attracts and gains the trust of new clients and consumers.

2 Ethics and Aesthetics in Organizations to Attract Talent Through AI

In the nineteenth century, Peirce, the father of pragmatism, enunciated three 'normative sciences', which deal with the self-controlled or deliberate, that which is oriented to an end. They are Aesthetics, Ethics, and Logic. He proposed Aesthetics as the science of ideals, of what is objectively admirable, Ethics as the science of what can be considered right or wrong, and Logic as the science of the general laws of signs, associated with the truth or validity of reasoning (CP 2.199).

From this perspective, the question arises: How do normative sciences relate to the everyday life of organizations? In principle, aesthetics can be equated with strategies because they represent what one wants to achieve, because they are admired. Ethics is applied to the evaluation of the moral consequences—internal and external—of organizations' actions and decisions. And Logic can guide judgments of whether, and how, an organization's guiding purposes are expressed in the actions of its managers.

When considering AI in organizations, it is worth questioning the implications of linking aesthetics with ethics. A first answer is that the linkage "...should help to prevent the passion for innovation, a sentiment shared by artists and entrepreneurs, from getting out of hand and becoming so polarised that it self-destructs. Because innovation, in itself, is always of interest, but it loses its meaning if it is not directed towards a specific purpose, if it is purely arbitrary or gratuitous" (Sanchez Asiain, 1991, p.45).

A second response is that AI can also act as a catalyst for the beauty and attractiveness of organizations, when its application is oriented in that direction. So, beyond the technological and economic benefits that AI can bring as a consequence of its application, it can also bring intangible benefits such as reduced stress at work.

Thirdly, and conversely, just as AI methods and tools can benefit the organization, they can also be a source of uncertainty if they are seen as threats to jobs or workers' integrity. This chapter aims to explore people's perceptions of this potential dilemma.

Aesthetics, ethics, and logic are ideally linked, and coherence between them is to be expected in an organization. In everyday life, there is a close relationship between the methods used to program actions and the effects they produce. In addition, the particular history of each organization will strongly affect the degree to which they can use AI methods and tools.

The research described in the following section begins with an inquiry into what AI methods and tools the respondents know and which, if any, they apply. Then, we explore the strategies (Aesthetics) identified as those that most enhance the use of AI in organizations, the benefits or achievements (Ethics) that are obtained as a result of its use, and the main purposes (Logic) that justify the implementation of AI methods and tools.

The concept of AI has controversial aspects, for example, whether it is possible to consider something "artificial" as "intelligent". However, AI is present in everyday life, in such ways as autonomous vehicles used in different industries, in aviation systems with "automatic" functions, in facial recognition in security cameras, and in voice identification in mobile devices. It can be argued that AI is a major characteristic of the technological era in which we live. Without confusing it with organizational aesthetics, the beauty industry has developed "personalized" makeup suitable for the color of each skin, something like wall paints, which can be formulated to the consumer's taste, and "digital" cosmetics. Of course, there are other applications that relate to aesthetic medicine, such as skin scanning for therapeutic effects.

Beyond the advantages of AI, questions have been asked about several of its problematic aspects. These include the risk that automation poses a future largely managed by machines (Lucífora, 2023), the extent to which self-organization can be incorporated into machines, i.e., spontaneity (Olezza, 2023), and the ways in which automatic systems continue to encompass spheres with human and autonomous functions, such that AI might design human beings and decided what and how we will be (Torres Lozano, 2020). Some of these issues are evidenced in surveys on AI applied to OM in organizations.

3 AI Methodologies Predominant in Operations Management

The last few years have seen a technological revolution, the so-called Industrial Revolution 5.0, which has enhanced the importance of Artificial Intelligence (AI). AI is one of the technologies that involve, replicate, and/or imitate reasoning, processes, or actions that until now were considered possible only by human beings. Many people would have struggled to believe 20 years ago that by today, the first automated intelligent cars would already be in use, home automation and medical applications of AI would be developed, high-end robotics would exist, there would be voice assistants for cell phones, AI traffic management systems, and machines able to maintain themselves via intelligent networks. But what used to be a futuristic vision is already a reality in today's society and in the daily lives of people and organizations. There is still a long way to go: the emergence of AI will change society, but not necessarily so soon.

Following Hintze (2016), four types of AI are empowering the new reality:

- **Reactive AI.** Within AI, there are various difficulties, across a range of levels. Reactive machines are the most basic systems since they have little capacity to store memories or take advantage of past experiences to make decisions based on previous events. Machines react to what they see in the world; they do not analyze what is inside them. That is, they are generally computers that process and make decisions about specific tasks assigned to them. Examples include Google's Alpha Go or IBM's Deep Blue. Deep Blue is a chess computer that can identify the pieces on a chessboard, reason the moves of each piece, and establish the most optimal moves among a multitude of possibilities (Hintze, 2016).
- **Memory-based AI.** This refers to machines programmed to recapitulate and go back to the past. However, the information about the past is transient and cannot be stored as part of the machine's experience. An example is that of driverless cars, which can be programmed with a series of events (stopping at a crosswalk, arrival at a traffic light, identification of an obstacle in the lane, speed measurement, identification of other cars in the opposite direction, among others), and can make the right decision. But it cannot go beyond that: this information cannot be transferred into a process of learning from an experience and improved in the next event (Hintze, 2016).
- **Deep Learning-based AI** is an advance on the use of limited memory. It not only configures the machine to act on a repetitive event but also tries to understand human behavior. The main driver of this type of AI is to understand that humans have emotions, feelings, and thoughts about how they like to be treated. According to Hintze (2016), this adds difficulty to its programming, which is based on adjusting to how people react to the decisions taken).
- **Knowledge-based AI** is the latest type of AI available to date. It consists of building a system capable of creating representations of oneself. That is, machines that have consciousness (have feelings, thoughts, moods), not just an understanding of human consciousness (Hintze, 2016).

This new technology requires the rapid transformation of ideas, products, or services, and their constant updating. It also necessitates the combination of new and current ideas, with a visionary attitude. It faces a constant challenge to be fashionable and to maintain its position as a pioneer (Agrawal et al., 2023).

In general terms, AI uses a wide range of tools to collect, analyze, and test data, which will later provide relevant and useful information. Some of the methodologies that comprise AI are Robotics, Generative AI, Machine Learning, Fuzzy Logic, Neural Networks-based AI, Super Artificial Intelligence, Expert Knowledge-based AI, and Bayes Networks-based AI (Allawi et al., 2018).

3.1 Robotics AI

The evolution of robots goes back many years, with four generations already in existence, and a fifth generation coming soon. In 1969, Victor Scheinman developed the first robotic arms, known as pick-and-place manipulator robots. First-generation robots serve as support resources for industrial machines. In other words, they require constant human supervision in the production process. In the 1980s, the second generation emerged, taking the programming of robots a step further to give them the ability to learn. This generation uses tactile sensors, proximity pressure, vision systems, radar, and sound, involving feedback, and needs only occasional, not constant human supervision. The main feature of the third generation of robots, from the late 80s and 90s, is that they have sensorized control; they are also known as AGV (Automatic Guided Vehicles). Here, the sensors are tactile and visual, and allow control by means of servomechanisms of self-guided movement and pressure (Chen & De Luca, 2021). Third-generation robots are programmable and autonomous, and require almost no supervision since they have their own defined controller. Finally, the fourth generation robots, developed in the early 2000s, are known as 'intelligent': they can adapt to the environment, learn from it, and make decisions immediately. They incorporate the Internet of Things (IoT) and, for the first time, AI (Geraci, 2010).

This is when, for example, home automation is developed, able to perform complex construction, cleaning or maintenance tasks. However, given the speed with which technologies evolve, engineers and scientists are already discussing a fifth generation. This new generation predominates because robots are unique. This means that robots have full artificial intelligence, equal or even superior to human intelligence (Torresen, 2018).

3.2 Machine-Learning AI

Machine Learning (ML) or automatic learning is another of the methodologies applied by AI that allows machines to learn without necessarily being programmed, i.e., they learn from the data they collect. ML can identify patterns among an infinite amount of data to estimate predictions and deduce the most optimal solution for a problem. It is often used in similarity detection or behavior prediction (Tyagi & Chahal, 2022). The major difference over previous methodologies is that ML can adapt instantly to the data as it is integrated into the system.

3.3 *Fuzzy Logic AI*

Fuzzy Logic (Zadeh, 1976) is an AI methodology for dealing with imprecise information. It uses a series of expressions that are neither entirely true nor false. In using Fuzzy Logic, when you have a series of data to make a decision, you cannot start from all or nothing, black or white, but you identify an intermediate point—or a definition of ranges—that allows you to make one decision or another (Kreinovich & Mukaidono, 2000). Fuzzy Logic is used for concepts that acquire a truth value within a set of values that oscillate between an absolute truth and a total falsehood. An example is the temperature range of a thermostat.

3.4 *Neural Network AI*

Neural Networks—also referred to as Artificial Neural Networks (ANN) or Simulated Neural Networks (SNN)—date back many years. In 1943, Warren S. McCulloch and Walter Pitts were already studying brain cells connected to neurons. In 1989, Yann LeCun studied the architecture of neural networks and their use for training algorithms.

Neural Networks are a subset of Machine Learning and are at the heart of Deep Learning. The way they work tries to simulate the human brain and mimics the way biological neurons signal to each other. Neural Networks are made up of nodes or artificial neurons, replacing brain neurons, that are interconnected with each other, and each of them has a weight. The fundamental goal of Neural Networks is to train algorithms to learn, and thereby improve the accuracy of decision-making over time (Abghari et al., 2012). Once the nodes learn and adjust themselves scrupulously, they form a very powerful computational tool for quickly classifying and clustering data.

3.5 *Artificial Superintelligence: ASI*

Another methodology that goes further in AI is currently being developed: artificial superintelligence (ASI). Practical literature is scarce at the moment. However, its theoretical concept is beginning to be developed in a possible future scenario. Artificial superintelligence is posited as capable of surpassing human intelligence, both in knowledge and skills (Bennett, 2023). It is a new technique that scares many experts and researchers, since there is a range of ethical issues, including but not exclusively those involving safety, autonomy, and human values.

3.6 Expert Systems AI

Some researchers are currently attempting to develop AI capable of making decisions and recommendations as if it were a specialist. Expert systems applied in AI make it possible to simulate the human ability to deal with specific problems and make complex decisions. To date, its use is limited, as it is in the process of development, although it is already beginning to be used, for example, in medicine, in the identification of fraud in banking, and in the development of engineering (Burnashev et al., 2020).

Expert systems are computer programs designed to simulate the analytical skills and knowledge of a specialist in a specific area (Krishnamoorthy & Rajeev, 2018). Developed using a series of algorithms, they tackle difficult questions and offer users recommendations, decisions, and solutions. Starting in the 1990s, fuzzy logic began to be used, approaching human reasoning, and today's systems have been fostered by machine learning.

3.7 Bayesian Network AI

Another of the methodologies emerging from AI based on expert knowledge is based on the work of Judea Pearl's (1988) invention of Bayesian Networks. They aim to define complex probabilistic models, and were a revolution for AI. They simplify models, making them easier to understand, taking into account only the most significant interactions and identifying the existence of errors. Bayesian networks are expert systems that present probabilities based on uncertain knowledge (Scanagatta et al., 2019).

Bayesian networks have succeeded in developing a language that allows us to use human knowledge about the rest of the world consistently. Even in the COVID-19 pandemic, it was applied to making inferences in the health domain and in control decisions (Aliyu et al., 2021).

4 Research on Artificial Intelligence Applied to Operations Management

This chapter identifies the best-known AI methods and tools, those applied daily in organizations, and examines ways of assessing the strategies used and the effects of their application in Operations Management (OM). In order to meet these objectives, the following research work was carried out.

4.1 Sample

One hundred professionals were surveyed, operations managers of different organizational processes. At the time of the study, only five of them were not working. 52% reported as male, 47% female and 1% preferred not to answer. 24% were 17–27 years old; 35% 28–42; 26% 43–57; and 15% were over 58.

The highest percentage, 42%, have worked for the same organization for more than 10 years; 34% have worked there for between 1 and 10 years, and 20% for less than a year. The respondents work in various fields of activity, with the highest frequency, 23%, dedicated to management and consulting services, followed by 22% dedicated to education and research. Most (54%) work in medium and large companies, with between 501 and more than 1000 employees; most have graduate or postgraduate qualifications, and around 50% hold an intermediate position in their organizational structure.

4.2 Questionnaire

A questionnaire called "Artificial Intelligence. Applied to Operations Management" was administered. Its purpose was to identify the AI tools known and used in the professional field and to elicit the respondents' assessments of the qualities and attractiveness of the different AI methods and tools.

The MAIA method (Sastre, 2016) was used to elaborate the questionnaire, a tool that allows us to diagnose the meanings that people, or groups of people, attribute to complex phenomena or concepts.

Table 1 shows the dimensions and the propositions included in the questionnaire.

The questionnaire has 39 variables or propositions to be evaluated by respondents on a scale of 1–10 (1 the least valued and 10 the most valued). The answers were then grouped, and the results were analyzed using both qualitative and quantitative methods.

Table 1 Dimensions of the MAIA method analyzed in OM AI

Dimensions	Methods	Actions	Strategies (Aesthetic)	Evaluations (Ethic)	Purposes (Logic)
Propositions (Quantity)	11	5	6	8	9

Source: Table prepared by the authors based on the MAIA method of Sastre (2016)

4.3 Analysis of Results

4.3.1 Methods and Tools They Know and/or Use (Quadrant 2)

Twenty-seven people said they did not know any of the Artificial Intelligence methods listed in Sect. 3 of this paper. An attempt was made to attribute this fact to factors such as age, level of education, or the position they hold in the organizations where they work, but nothing in particular was found to identify them as a homogeneous group. The other 73 respondents rated Artificial Intelligence methods with varying scores based on what they consider most suitable for use in OM in organizations. A score of 1 was given to the methods they do not know; as seen in Table 2, many are unknown to most people. The best known, by 85% of the sample and the most valued for application in OM, is Machine Learning, followed by Knowledge-Based AI.

Another section of the questionnaire asks about the tools that respondents know and use. People probably know about AI more by the name of the tools than by how they were generated. The result of this inquiry includes the entire sample, i.e., it includes the 27% who in the previous question answered that they do not know any AI method. Table 3 shows that the percentage of those who "do not know any AI method or tool" was reduced to 12% of the respondents.

Conversational chats, considered personal assistants, are the best known. In first place is GPT Chat (Open AI), followed by others such as Siri (Apple), BingChat (Microsoft), Bard (Google) and ErnieBot (Baidu).

Table 3 shows that 11% of the respondents know between one and three tools besides Chatbots and GPT Chats. These include tools for pricing generators (Pricing); sales and CRM assistance (Salesforce Einstein); text generators (Copy.

Table 2 Known and valued AI methods

Evaluation in number of persons	Unknown (1 point)	Little use in OM (2–5 points)	Average profit in OM (6–8 points)	High usefulness in OM (9–10 points)
Reactive AI	35	4	21	13
Memory-Based AI	24	2	23	24
Deep Learning-Based AI	18	3	19	33
Knowledge-Based AI	17	2	20	34
Machine-Learning AI	11	4	13	45
Neural Network AI	28	4	15	26
Fuzzy Logic AI	36	5	16	16
Expert Systems AI	22	3	16	32
Robotics AI	20	3	18	32
Bayesian Network AI	43	1	17	12
Artificial Superintelligence ASI	42	2	14	15

Source: Table prepared by the authors based on $n = 100$

Table 3 Known and used tools

Tools	Percentage known	Percentage used
None	12%	45%
Only Chat GPT	42%	24%
Chat GPT and 1 different Chatbots tool	11%	6%
Chat GPT and between 2 and 3 different Chatbots tools	11%	6%
Chat GPT and more than 3 different tools from Chatbots	6%	1%
Chat GPT and Other Chatbots	7%	6%
Types of IA	4%	2%
Others	7%	10%
Total	100%	100%

Source: Table prepared by the authors based on $n = 100$

AI, Jasper, Frase); image generators (MidJourney, Dalle-E, Picsart); video generators (Lumen %); presentation and graphics creators (Beautiful AI, Canva, Illustrator) and search engines (Google, You, Bing). In all cases, the known tool is related to the position and type of OM tasks executed in the organization where the respondent works. A strikingly high percentage (69%) do not use any tool, or only use ChatGPT.

4.3.2 Overall Averages (Table 4)

4.3.3 Segmentations

In general, a sample is segmented in order to verify the association between different variables. The selection of the independent variables is based on prior knowledge, both theoretical and practical, which makes it possible to construct a series of hypotheses that predict different responses in each grouping.

The first segmentation is made by age, since young and older adults are expected to display different behaviors vis-a-vis AI.

As Table 5 shows, in the dimension that asks about the most effective actions and decisions to promote the use of AI, people between 17 and 42 years of age consider "Investment in technology" to be the most effective, while people over 43 years of age point to "Training of personnel" as the most effective.

In the case of the strategies that are most enhanced using AI, people aged over 58 years value "Competitive" strategies the most, while the rest of the sample considers that "Change and innovation" strategies are the most enhanced.

The dimension that addresses the benefits of using AI in the OM, people over 58 consider "Improved information processing" and "Speed in decision making" as the most valued achievements, with the same average score. People between 17 and 57 valued "Optimization of time" most highly.

Table 4 Most valued propositions

Dimensions	Description	Proposition
Methods	High usefulness in OM	Machine-Learning AI
Actions	Most effective in promoting the use of AI in OM	Investment in technology
Strategies	Most enhanced by the use of AI	Change and innovation strategies, because it facilitates the customization of products and/or services
Evaluations	The value of AI applied to OM in terms of the effects achieved by its use	Optimization of time
Purposes	The rationale for the implementation of AI methods and tools in OM	To increase productivity

Source: Table prepared by the authors based on $n = 100$

Table 5 Segmentation by age

	Age			
Dimensions	17–27	28–42	43–57	>58
Actions	Investment in Technology	Investment in Technology	***Training of personnel***	***Training of personnel***
Strategies	Change and Innovation	Change and Innovation	Change and Innovation	***Competitive***
Evaluations	Optimization of time	Optimization of time	Optimization of time	***Improved information processing & Speed in decision-making***
Purposes	Increase productivity	Increase productivity	***Increase profitability & Reduce operating costs***	Increase productivity

Source: Table prepared by the authors based on $n = 100$

The purposes justifying the implementation of AI methods and tools in the OM of organizations also show divergence in the answers. Most of the sample considers that "To increase productivity" is the reason that most justifies its implementation. In turn, the segment between 43 and 57 years of age pointed to "To reduce operating costs" and "To increase profitability".

The second segmentation by educational level showed different associations, as seen in Table 6.

Table 6 shows that the associations made vary by educational level.

Table 6 Segmentation by educational level

Dimensions	Level Completed Post graduation	Uncompleted Post graduation	Completed University studies	Uncompleted University studies
Actions	Training of personnel	Training of personnel	Training of personnel	***Investment in Technology***
Strategies	Change and Innovation	Change and Innovation	***Anticipation and problem resolution***	Change and Innovation
Evaluations	Optimization of time	Optimization of time	Optimization of time	Optimization of time
Purposes	Increase productivity	***Increase profitability***	Increase productivity	Increase productivity

Source: Table prepared by the authors based on $n = 100$

4.3.4 Correlations

Correlation analysis seeks to corroborate relationships between quantitative variables. Pearson's correlation index is used as an indicator in this research, since the sample size is significant according to the formula of Streiner and Norman (2015). The strength and direction of the correlation vary between −1 and 1; an r correlation between ± 0.5 < ± 0.7 can be considered high, and between ± 0.7 < ± 1 is very high.

The first very high relationship between the variables establishes that staff retention is one strategy that is most enhanced using AI, because it makes organizations more attractive. This strategy correlates strongly (r = 0.723) with staff satisfaction and loyalty.

The second very high correlation involves data warehousing and IT security (r = 0.716).

The third very high correlation links customer satisfaction with the achievement of customer and consumer trust (r = 0.891). The fourth very high correlation links improvements in information processes with the optimization of time (r = 0.775).

The last very high correlation relates to purpose, i.e., the Logic quadrant in the MAIA method and associates cost reduction with increased productivity (r = 0.721) (Fig. 1).

Twenty-three correlations were found with r-factors between 0.5 and 0.7, considered 'high', which indicates that there is a close relationship between the propositions (variables) considered in the questionnaire.

5 Conclusions

It is undeniable that AI tools are valuable in practical life, both personally and professionally. They allow us to optimize the time spent on such routine tasks of searching and selecting whatever we need, be it images, videos, texts, or ideas.

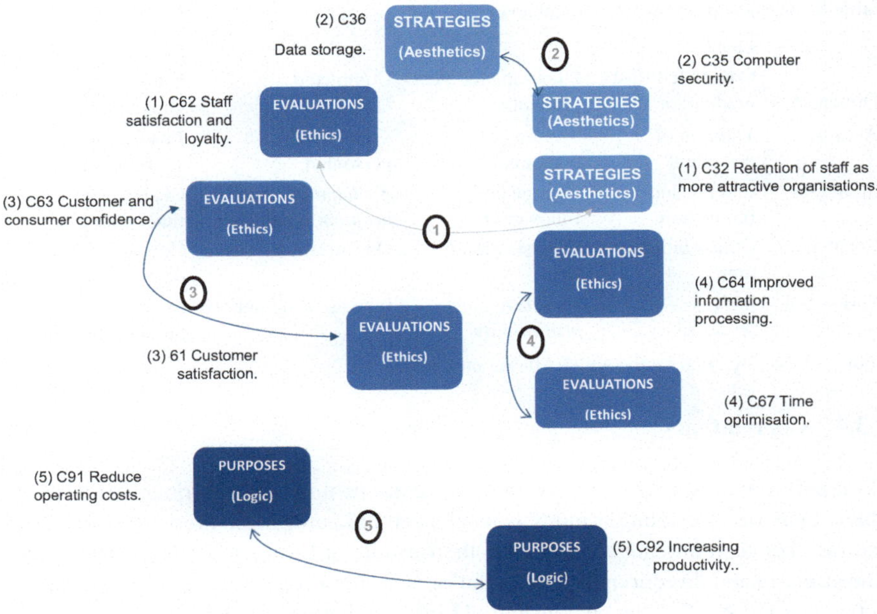

Fig. 1 Correlations: very high between 0.7 < 1.0

Currently, AI is based on artificial neural networks and deep learning algorithms, which are highly specialized systems designed to perform specific tasks. The extensive database they access provides answers that help with everyday tasks, although some experts warn about the reliability of the results of the queries. Tools such as ChatGPT and other Chatbots were designed and developed by companies with an eye to the market; they do not investigate the scientific accuracy of their answers. There is no discrimination between true and false.

In addition to powerful processors, data sets are needed to develop AI methods, and the search for them has become a real battle between development companies. Data protection, *in res* privacy or copyright, is an ethical dilemma that societies must face and regulate through their political systems. The question is whether AI is functional or not to society, remembering that it is not the tool itself that is harmful, but the use to which it is put.

Based on the assumption that the use of AI in the management of operations in organizations, the study investigates how strategies and effects of actions in the application of AI are assessed. They are simply indicators that point toward future lines of research.

So, our results—from a heterogeneous sample, with professionals with different levels of education, working in large and small companies with different branches of activity, and a variety of other characteristics—show that there is little knowledge of the AI methods and tools currently being developed. Likewise, the recognized use of these tools is limited. Given this lack of knowledge, one of the proposals made in this chapter is that more staff training should be provided to promote its use.

The correlation analysis revealed close associations between the variables referring to Aesthetics and Ethics. It confirmed that, for a group of randomly chosen managers, most of whom knew nothing of AI, the application of AI in OM favors the growth of organizations, as it allows them to attract new clients and consumers and gain their trust. However, despite the correlation found between staff satisfaction and loyalty because AI makes organizations more attractive, the results do not show that the use of AI tools is a catalyst for the beauty and attractiveness of organizations.

According to the results, the rationale for implementing AI methods in OMs is economic. AI is perceived as a valuable technology for increasing productivity and reducing operational costs. But it is not appreciated, for example, that another benefit of increasing the automation of processes is reduced stress at work.

Moreover, another proposition that does not follow a purely economic rationale—reducing disagreements among staff because AI tools provide answers that can contribute to consensus—was the least valued in the entire questionnaire. Finally, AI is not perceived as adding to the threat of replacing people with machines.

6 Implications and Future Lines of research

One practical implication of these findings is that AI should be studied in greater depth in the field of operations management, given its wide use and that there is still much to learn. This chapter serves as a starting point and as the incentive to continue researching and developing other studies. These would include the training of employees and managers in AI and how the tools currently being developed might be used in logistics. This chapter is of general interest to experts in the field of AI but also—especially—to those who are considering or beginning to implement it in their organizations, especially with regard to OM. The literature review showed that the tools currently being applied are becoming more and more advanced. But, as our findings show, real-world implementation is much delayed and can be strengthened, in OM and other departments. It may also interest professionals dedicated to logistics operations who intend to broaden their knowledge and apply new techniques.

One major recommendation is that more research investigates the extent of knowledge about AI in OM and the practicability of incorporating the former in the latter. Given that the sample used belongs to a specific sector and with similar characteristics, it would be valuable to explore whether our results are reflected among subordinates and younger middle managers.

Another possible line of future research would be to find the causes of the lack of use of AI tools: possibilities include that it is a socio-economic problem, that the advantages of AI have not been sufficiently disseminated, and the lack of economic resources.

Finally, it would be interesting to investigate knowledge and attitudes within universities, and not just in engineering or information systems courses. One could consider implementing training programs on the different types of AI and the tools currently being developed.

References

Abghari, H., Ahmadi, H., Besharat, S., & Rezaverdinejad, V. (2012). Prediction of daily pan evaporation using wavelet neural networks. *Water Resources Management, 26*, 3639–3652. https://doi.org/10.1007/s11269-012-0096-z

Agrawal, A. K., McHale, J., & Oettl, A. (2023). *Artificial intelligence and scientific discovery: A model of prioritized search* (No. w31558). National Bureau of Economic Research. https://doi.org/10.3386/w31558

Aliyu, S., Zakari, A. S., Adeyanju, I., & Ajoge, N. S. (2021). A Bayesian network model for the prognosis of the novel coronavirus (COVID-19). In *Computational science and its applications—ICCSA 2021: 21st international conference, Cagliari, Italy, September 13–16, 2021, Proceedings, Part IX 21* (pp. 127–140). Springer. https://doi.org/10.1007/978-3-030-87013-3_10

Allawi, M. F., Jaafar, O., Mohamad Hamzah, F., Abdullah, S. M. S., & El-Shafie, A. (2018). Review on applications of artificial intelligence methods for dam and reservoir-hydro-environment models. *Environmental Science and Pollution Research, 25*, 13446–13469. https://doi.org/10.1007/s11356-018-1867-8

Barrena, S. (2007). *La razón creativa. Crecimiento y finalidad del ser humano según C. S. Peirce*. Rialp.

Belhadi, A., Kamble, S., Jabbour, C. J. C., Gunasekaran, A., Ndubisi, N. O., & Venkatesh, M. (2021). Manufacturing and service supply chain resilience to the COVID-19 outbreak: Lessons learned from the automobile and airline industries. *Technological forecasting and social change, 163*, 120447. https://doi.org/10.1016/j.techfore.2020.120447

Bennett, M. T. (2023). Enactivism & objectively optimal super-intelligence. *arXiv Preprint*. arXiv:2302.00843. https://doi.org/10.48550/arXiv.2302.00843

Burnashev, R. A., Gabdrahmanov, R. G., Amer, I. F., Vakhitov, G. Z., & Enikeev, A. I. (2020). Research on the development of expert systems using artificial intelligence. In *Information systems architecture and technology: Proceedings of 40th anniversary international conference on information systems architecture and technology—ISAT 2019: Part II* (pp. 233–242). Springer. https://doi.org/10.1007/978-3-030-30604-5_21

Chen, Y., & De Luca, G. (2021). Technologies supporting artificial intelligence and robotics application development. *Journal of Artificial Intelligence and Technology, 1*(1), 1–8. https://doi.org/10.37965/jait.2020.0065

Dankan Gowda, V., Sridhara, S. B., Naveen, K. B., Ramesha, M., & Naveena Pai, G. (2020). Internet of things: Internet revolution, impact, technology road map and features. *Advances in Mathematics: Scientific Journal, 9*(7), 4405–4414. https://doi.org/10.37418/amsj.9.7

Fosso Wamba, S., Queiroz, M. M., & Braganza, A. (2022). Preface: artificial intelligence in operations management. *Annals of Operations Research, 308*, 1–6. https://doi.org/10.1007/s10479-021-04450-0

Geraci, R. M. (2010). *Apocalyptic AI: Visions of heaven in robotics, artificial intelligence, and virtual reality*. Oxford University Press.

Hintze, A. (2016). Understanding the four types of artificial intelligence. *Consulté le, 10*(11), 2021.

Kreinovich, V., & Mukaidono, M. (2000). Intervals (pairs of fuzzy values), triples, etc.: can we thus get an arbitrary ordering?. In *Ninth IEEE international conference on fuzzy systems.*

FUZZ-IEEE 2000 (Cat. No. 00CH37063) (Vol. 1, pp. 234–238). IEEE. https://doi.org/10.1109/FUZZY.2000.838664
Krishnamoorthy, C. S., & Rajeev, S. (2018). *Artificial intelligence and expert systems for engineers*. CRC Press. https://doi.org/10.1201/9781315137773
LeCun, Y. (1989). Generalization and network design strategies. *Connectionism in Perspective, 19*(143–155), 18.
Lucífora, M. C. (2023). El impacto de la Inteligencia Artificial en la semiosis humana desde la teoría de Peirce. In *IX Jornadas de Peirce en Argentina* (pp. 70–77). Academia Nacional de Ciencias.
McCulloch, W. S., & Pitts, W. (1943). A logical calculus of the ideas immanent in nervous activity. *The Bulletin of Mathematical Biophysics, 5*, 115–133. https://doi.org/10.1007/BF02478259
Olezza, M. (2023). Inteligencia Artificial, semiosis y sistemas complejos. In *IX Jornadas de Peirce en Argentina* (pp. 79–86). Academia Nacional de Ciencias.
Pearl, J. (1988). *Probabilistic reasoning in intelligent systems: networks of plausible inference*. Morgan Kaufmann.
Peirce, C. S. (1893/1931). *Collected papers of Charles Sanders Peirce*. D. Hartshorne & P. Weiss (Eds.). Harvard University Press.
Rifkin, J. (2014). *The zero marginal cost society: The internet of things, the collaborative commons, and the eclipse of capitalism*. St. Martin's Press, Palgrave MacMillan.
Sanchez Asiain, J. A. (1991). *La empresa como realidad estética* (pp. 9–56). Real Academia de Bellas Artes de San Fernando. Discurso leído el 4 de abril e impreso en Madrid.
Sastre, R. (2016). An analytical methodology for management research: Insights from semiotics. *Studia Oeconomica Posnaniensia, 4*(5), 199–218. https://doi.org/10.18559/SOEP.2016.5.12
Scanagatta, M., Salmerón, A., & Stella, F. (2019). A survey on Bayesian network structure learning from data. *Progress in Artificial Intelligence, 8*, 425–439. https://doi.org/10.1007/s13748-019-00194-y
Streiner, D. L., & Norman, G.-R. (2015). *Health measurement scales: A practical guide to their development and use*. Oxford University Press. https://doi.org/10.1093/med/9780199685219.001.0001
Torres Lozano, D. (2020). Ethical erosion and the handling with artificial intelligence. *Logos, 134*, 43–58. https://doi.org/10.26457/lrf.v0i134.2527
Torresen, J. (2018). A review of future and ethical perspectives of robotics and AI. *Frontiers in Robotics and AI, 4*, 75. https://doi.org/10.3389/frobt.2017.00075
Turing, A. M. (1936). On computable numbers, with an application to the Entscheidungsproblem. *Journal of Mathematics, 58*, 345–363.
Tyagi, A. K., & Chahal, P. (2022). Artificial intelligence and machine learning algorithms. In *Research anthology on machine learning techniques, methods, and applications* (pp. 421–446). IGI Global. https://doi.org/10.4018/978-1-6684-6291-1.ch024
Yang, C., Feng, Y., & Whinston, A. (2022). Dynamic pricing and information disclosure for fresh produce: An artificial intelligence approach. *Production and Operations Management, 31*(1), 155–171. https://doi.org/10.1111/poms.13525
Zadeh, L. A. (1976). A fuzzy-algorithmic approach to the definition of complex or imprecise concepts. *International Journal of Man-Machine Studies, 8*(3), 249–291. https://doi.org/10.1016/S0020-7373(76)80001-6

The Impact of Artificial Intelligence on HR Practices

Sangeeta Nar, Kerstin Rego, Christian Scharff, and Andreas M. Hilger

Abstract The growing adoption of generative Artificial Intelligence (AI) has sparked significant interest in the potential uses of AI in the context of organizational processes and practices. In this chapter, we review the available literature on the use of AI in HR practices with two main objectives in mind: to assess the status quo of the academic literature, and to highlight the HR practices that are most widely studied. Our findings show that the inconsistent application of the term "artificial intelligence" impacts the ability to draw generalizable conclusions. The literature also suggests that the HR practices most likely to be affected by AI are those related to personnel recruiting and employee relations. We argue that given the nature of AI, the application potential also extends to other HR practices, such as those related to HR controlling, which remain underexplored.

1 Introduction

Artificial intelligence applications are developing rapidly and becoming increasingly embedded in society. At the same time, the emergence of generative AI is attracting widespread media and corporate attention. This progress is increasingly impacting human resources (HR) departments (Oracle Corporation, 2022).

According to HR practitioners and consultancy firms, companies are turning to AI for employee records management, payroll processing, benefits administration, performance management, recruitment, hiring, and onboarding of new employees

S. Nar (✉)
Chair of Innovation and Technology Management, University of Regensburg, Regensburg, Germany
e-mail: sangeeta.nar@ur.de

K. Rego · C. Scharff · A. M. Hilger
Chair of Leadership and Organization, University of Regensburg, Regensburg, Germany
e-mail: kerstin.rego@ur.de; christian.scharff@ur.de; andreas.hilger@ur.de

M. T. Del Val Núñez et al. (eds.), *Artificial Intelligence and Business Transformation*, Contributions to Management Science,
https://doi.org/10.1007/978-3-031-58704-7_9

(McKinsey & Company, 2023). AI is employed, for instance, to streamline recruitment, onboarding and training, and, more specifically, for employee analytics and to redesign tasks to increase efficiency or to determine employee schedules (Oracle Corporation, 2019; Cappelli & Rogovsky, 2023). In anticipation of efficiency benefits such as reduced bias, corporate AI implementation in HR is expected to grow in the coming years. Some advocates even anticipate a better employee experience (Chevalier, 2023), partly due to growing acceptance of AI. For example, two-thirds of respondents surveyed by Oracle stated that they would trust a robot over their manager and would enjoy working with AI robots (Oracle Corporation, 2019).

However, some critics caution against the use of AI, pointing to potential detrimental impact on the workforce and HR, such as deskilling, the possibility of job loss and issues related to accountability, morality, ethics, and acceptance (Cappelli & Rogovsky, 2023). From a managerial perspective, there are also concerns about when it makes sense to use AI to perform tasks better than humans and when its use is counterproductive.

In this chapter, we analyze extant literature to determine how AI changes HR practice and potential risks of using AI in HR. We present the current state of knowledge on the introduction of AI in HR and investigate measurable and demonstrated effects of AI on HR practices, focusing on the practices of strategic HR, recruitment and selection, training, development, and talent management, as well as employee relations. Based on our findings, we present implications for the future of HR research and practice, particularly concerning the implementation of AI and the impact of human involvement in human resource tasks.

2 Implementing AI in HR

As the interest in AI in HR has grown, so has the academic interest in the topic (e.g., Jatoba et al., 2023). Two research directions can be distinguished: first, how AI replaces HR as a department in algorithm-based companies such as platform companies (e.g., Waldkirch et al., 2021), and second, how AI supports HR in 'classic' corporate forms (e.g., Jatoba et al., 2023). In this chapter we focus on the second perspective. Particular attention will be paid to studies that empirically examine the connection between the implementation of AI-based technology and HR functions.

There are a number of different views on what exactly AI is. However, common definitions of AI share one or more of the following characteristics: Artificial intelligence has been defined as "shorthand language for a set of complex algorithms" (Moser et al., 2022). Others highlight the aspect of AI taking on tasks that previously required complex human mental effort (e.g. Raisch & Krakowski, 2021), while artificial intelligence is also defined as a technology that is capable of independent learning (Burrell, 2016; Holton & Boyd, 2021).

AI implementation in HR typically occurs in two stages: The first stage leads to the automation of the most time-consuming HR processes while the second stage involves a more strategic shift, in which AI works in collaboration with employees (Altemeyer, 2019).

The responsibilities of the HR department can be summarized either according to the roles that the department fulfils or according to the practices that take place within the department. To address the changing demands in HR, Ulrich (1996) defined four HR roles and noted that rapidly changing technology poses one of the greatest challenges to HR departments, requiring a rethinking of HR as an organizational unit. As a result, this would require a redefinition of the HR practices, functions, and roles (Ulrich, 1996). This approach allows HR to be understood not in terms of the processes that take place in the department, but in terms of the outcomes that are the responsibility of the department. According to this view, HR must fulfil four distinct roles: Strategic Partner, Administrative Expert, Employee Champion, and Change Agent (Ulrich, 1996).

The first role of the Strategic Partner involves aligning business strategy with the HR strategy and translating strategy into HR practices. The second role embedded in HR is that of the Change Agent. This role is responsible for facilitating strategic transformation and enabling change capability. Both the Strategic Partner and the Change Agent possess a high degree of strategic orientation. Conversely, the roles of Employee Champion, who supports employees in their daily business, and Administrative Expert, who is responsible for designing and improving HR processes, are more operational in nature (Ulrich, 1996).

As defined by Ulrich (1986, p. 173), "Human resource practices represent the policies, procedures, systems, and activities used to shape, monitor, and direct attention of people within the organization". Examples of HR practices include Strategic HR practices, Recruitment and Selection practices as well as practices related to Training, Development, and Talent Management. Strategic HR practices, for example, align HR with business objectives in such a way that the changing needs of the business are supported with appropriate, effective human resources. These practices cover wide-ranging areas, such as human resource information systems strategy and organizational design issues. Recruitment and Selection practices are implemented to appropriately source and hire new talent. HR Controlling is responsible for collecting, analyzing, and managing employee-related data, including productivity, headcount levels, and span of control. Training, Development, and Talent management focus on employee growth through feedback sessions, career path planning, and relevant developmental opportunities. Employee Relations practices provide daily support to employees. This includes assistance with personnel or company policy issues and compliance matters. Research has shown that human resource practices have the potential to improve business performance by increasing productivity and deliver financial results (Huselid, 1995). This suggests that organizations stand to benefit significantly from implementing effective HR strategies.

3 Exploring the Impact of AI on HR Practice

In this section, we draw on empirical studies to explore AI-driven changes in specific HR practices and associated benefits and risks.

3.1 Strategic HR Practices

As the strategic relevance of HR has been the subject of much discussion and is often recognized as the most important HR role in contemporary organisations, we begin by considering the potential use of AI in this context. While there is currently a lack of empirical evidence on the use of AI in strategic HR, a number of different potential deployment options are generally plausible. For example, it stands to reason that AI's ability to perform routine tasks will release resources that can be devoted to more strategic HR activities (Malik et al., 2022; Sithambaram & Tajudeen, 2023). Thus, the role of HR in organisational development is likely to change as well. Accordingly, the current research suggests that AI only provides benefits for the strategic practices of HR while risks are not reported.

3.2 Recruitment and Selection Practices

One of the most discussed HR practices related to AI implementation is the recruitment and placement of personnel, which can be supported by bots, and virtual, digital, or personal assistants (Malik et al., 2022). AI can create job postings and advertisements, in addition to screening and scheduling interviews, and can even generate offer letters (Niehueser & Boak, 2020). Furthermore, it can reduce the need for in-person on-campus recruiting and enable simultaneous recruiting at multiple universities, while the recruiters themselves focus on verifying the accuracy of AI-based assessments (Li et al., 2021). This could free the HR department from time-consuming routine activities, allowing more time for complex and problem-solving tasks (Malik et al., 2022; Sithambaram & Tajudeen, 2023).

Furthermore, the quality of recruiting can be improved by mapping internal cultural values and performance indicators against candidate qualifications and ability, and by including additional indicators of job fit. For example, measuring and assessing indicators such as reaction times and concentration levels for fleet transport drivers can contribute to making better recruiting decisions (Altemeyer, 2019). Furthermore, AI can support fast and efficient identification of candidate information, update new search terms based on changing job requirements, and help recruiters to understand changing recruitment needs (Li et al., 2021). In the assessment phase, AI could decrease biases and drive inclusivity by enabling access to a larger pool of potential candidates (Li et al., 2021). For example, Hofeditz et al. (2022)

found that candidate recommendations based on Explainable Artificial Intelligence (XAI) make selection processes more objective. The positive effects of AI implementation, however, have been shown to hinge on the performance capacity of the implemented technology and its ease of use (Niehueser & Boak, 2020).

Of all benefits of AI implementation in recruitment and selection, the increase of effectiveness is the most prominent. Particular mention is made of HR cost-effectiveness due to a reduced headcount, a higher value-added for business, and increased agility (Malik et al., 2022). In some cases, the length of the recruitment process is reported to be significantly reduced, for example, by lowering the time from identifying candidates to interviewing them from two weeks to three days (Altemeyer, 2019), or even more significantly, from weeks to minutes in candidate identification as stated by Niehueser and Boak (2020). However, not only does the process become more efficient in times of shortage of qualified labor (Li et al., 2021), but it could also improve in quality by becoming more objective and reducing discrimination against older and female candidates (Hofeditz et al., 2022) or increasing efficiency by reducing nepotism and favouritism (Kshetri, 2020). AI thus has the potential to mitigate recruiter bias and increase the diversity of potential candidates by including more atypical applicants (Li et al., 2021). Kshetri (2020) found that this enables a wider pool of candidates to be considered in the global south. Malik et al. (2022) report an increased personalization of services, which can lead to higher employee satisfaction and commitment. For example, AI enables companies to include factors such as empathy or emotional intelligence and may be used to analyze and maximize employee retention to reduce sunk costs associated with bad hires (Altemeyer, 2019). The increased efficiency could result in a reduction of HR headcount unless HR employees are upskilled for more value-adding jobs. However, this creates a need for continuous learning to work with AI or to apply higher-level skills (Sithambaram & Tajudeen, 2023) and the short life span of software may require continual relearning (Li et al., 2021). HR professionals skilled in human and data science will also be able to predict the impact of attrition on business, eliminating the need to constantly replace and recruit talent. For example, HR can leverage data to identify employees who can or cannot transition to new skills in times of change (Altemeyer, 2019). However, if the software is not updated frequently or secured properly, systems can be hacked, which may lead to financial risks, legal battles, and a loss in candidate and customer reputation (Sithambaram & Tajudeen, 2023). Depending on the regulatory context, the usage of AI for HR assessment may require legal approval and requires lengthy co-development, sharing of sensitive data with software providers for customization, and extensive integration work (Li et al., 2021).

In addition, AI may not deliver the expected results if it is not able to consider context. This is linked to a lack of precise control over sourcing results and to potentially overlooking promising candidates, who may not be suitable for roles where transferable skills are required. The quality of selection outcomes also relies strongly on the candidates' willingness to share large amounts of personal data and their technological ability, which may lead to challenges related to assessment completion and a need to nudge candidates to complete the process. This creates a trade-off between speed and effectiveness on the one hand and data quality and decision outcome on the other. The ability and willingness to engage on the part of a candidate may fluctuate depending on the candidate's age, which may exclude older but highly suitable candidates. Additionally, traits, speech patterns, and personalities may not be represented in the data and thus would not be detectable through AI. Resulting low-quality selection decisions, however, may lead to a decrease in the confidence recruiters have in AI, which in turn is reliant on the quality of the underlying data, e.g., when position titles are not standardized or when data sources are subject to frequent change such as social recruiting platforms (Li et al., 2021).

Overall, the predictive capability of assessment software as well as its return on investment rely on high-volume scenarios, which requires relying on the scores as humans cannot handle such volumes. Thus, the candidates that fall within the lower result tiers are not considered, which opens up the possibility of legal liabilities and a negative public image if the biased and flawed suggestions are made (Li et al., 2021). Consequently, discrimination may occur e.g., through lower selection of candidates of foreign race, as Hofeditz et al. (2022) have shown.

3.3 HR Controlling Practices

Given the ability of AI to process, analyze, and visualize large amounts of data, the deployment of artificial intelligence in HR Controlling could be taken for granted. However, present research provides little evidence of specific application scenarios. A notable exception is the study by Sithambaram and Tajudeen (2023), which surveyed twelve companies, only two of which indicated using AI for workforce planning optimization. It was also utilized to recommend improved time management to employees through the AI-based tracking of their productivity. Another significant benefit of AI for HR Controlling is the simplicity and speed with which it provides access to more data and information. However, the HR practices in HR Controlling are changing in that a thorough understanding of the underlying data is required so that corrections can be made to databases if necessary and HR Controlling can fully interpret the output (Sithambaram & Tajudeen, 2023).

Difficulties in measuring HR outcomes, such as establishing clear criteria of what makes a successful employee, may be a possible reason for the underutilization of AI in HR (Cappelli et al., 2020). For artificial intelligence to be effective, there needs to be a set of rules on which the AI can base its processing, and this is more difficult to achieve with HR-related metrics. In addition, the availability and quality of data related to HR metrics may not be at a point where the application of AI is reasonable given the associated costs and risks (Wirges & Neyer, 2023).

A potential threat is the risk of cyberattacks, which could explain the as yet untapped potential of using AI for HR controlling. As sensitive data such as payroll details and identity numbers are stored in HRIS systems, it is crucial to ensure data security before employing AI (Sithambaram & Tajudeen, 2023). Regulatory requirements create uncertainty in companies regarding the compliance of the collected workforce data with relevant standards (Shook et al., 2019). However, failure to comply with personal data protection regulations, which are unique to each country in which an organization operates, can result in significant penalties, as evidenced by the case of Interserve, a construction group that was fined £4.4 million in 2022 after falling victim to a phishing scam in 2020 (Omoigui, 2022).

3.4 Training, Development, and Talent Management Practices

Bots and assistants have the potential to support practices related to training, development, and talent management (Malik et al., 2022) and the changes in responsibilities between employees and artificial intelligence are expected to fundamentally change organisational development and training (Ardichvili, 2022).

Advanced training can be provided to employees to combat micro-tasking, which is a negative consequence associated with increased automation. Furthermore, establishing a learning culture that encourages self-directed learning and thus adaptability can facilitate the transition towards increased AI usage (Ardichvili, 2022).

The main benefit for HR is cost-effectiveness (Malik et al., 2022), as less staff is needed, more value can be added to the organisation, and HR agility is supported. Employees benefit as HR services are increasingly personalised, which leads to higher employee satisfaction and commitment (Malik et al., 2022). In addition, the automation of routine work brings with it the potential to shift the focus to more cognitively challenging tasks for employees throughout the organization.

However, the increasing use of AI is also associated with risks for HR practices. First, as electronic systems can be hacked, there is a need for higher cyber security. Therefore, HR professionals need to include an awareness of data security in their HR practices. The use of AI could also lead to an increase in menial and low-level tasks, such as data entry or reporting, which will necessitate the development of supplementary, job-enhancing activities (Ardichvili, 2022). As AI technology develops, so does the need for continuous learning for HR professionals (Sithambaram & Tajudeen, 2023). Furthermore, the development of expertise is challenged as there is a lack of exposure for workers to acquire and grow their expertise when tasks are increasingly performed by machines. This can result in deskilling, either through the loss of existing skills for senior professionals or through the lack of opportunities to grow new skills for younger professionals. In addition, Ardichvili (2022) argues that the use of AI impedes the development of expertise, as employees are less exposed to complete process and hence cannot easily grasp intuitively how minor changes could affect the process result (Ardichvili, 2022).

In conclusion, while AI offers great benefits for HR development, training, and talent management, it also poses risks due to an increasing dependency on the availability of technology.

3.5 Employee Relations Practices

Artificial Intelligence has numerous potential applications for Employee Relations, the function responsible for providing immediate support to employees and ensuring a positive relationship within the workforce. By automating routine tasks and releasing resources for more complex or value-added tasks, AI is changing these practices (Sithambaram & Tajudeen, 2023). One example of the implementation of AI is the use of chatbots for recurring inquiries about company or HR policies, such as leave policies, or for directing employees to relevant applications or forms (Sithambaram & Tajudeen, 2023). Furthermore, depending on the specific circumstances, AI has the potential to execute interpersonal tasks. For example, a study has shown that employees prefer to interact with AI rather than an HR professional when dealing with workplace harassment, particularly when employee satisfaction is low, and the work environment is perceived as hostile. As a result, AI may be used in select situations as a supplemental service to engage a broader range of employees (Bartosiak & Modlinski, 2022). Additionally, it can function as a decision support system for HR officers, leading to faster decision-making, for example, on disciplinary actions (De Obesso et al., 2023).

Nevertheless, there are risks associated with using AI in employee relations. Ensuring data security is critical to reducing the risk of hackers targeting sensitive data (Sithambaram & Tajudeen, 2023). Additionally, AI serving as a decision support system may influence HR professionals to make decisions that are potentially detrimental to employees compared to if they had made the decision without the influence of AI (De Obesso et al., 2023).

4 Implications for Research

We find an imbalance in research on HR practices, which limits the understanding of the impact of AI on HR. Compared to research focusing on the impact of AI on workforce adoption and behavior (see Basu et al., 2023; Del Giudice et al., 2023) and the future of work in general (see Poba-Nzaou et al., 2021), much less is known about the impact on the HR function and its practices. Additionally, there is a pronounced interest in the practices related to recruiting and talent management in the literature researching the use of AI in HR (Table 1). Due to the observational nature of the social sciences, practical implementation must precede empirical findings, as is the case with HR recruitment and selection.

In addition, literature contributions often do not disclose or clarify their understanding of AI (see Abbas et al., 2023; Altemeyer, 2019; Bartosiak & Modlinski, 2022; De Obesso et al., 2023; Kumar et al., 2019). Other authors explain numerous AI technologies but remain unclear about which type they have studied (see Ardichvili, 2022). This is detrimental because without critical information about the technology being reported on, readers cannot accurately categorize and compare

Table 1 Impact of artificial intelligence on HR practices

HR practice	Impact of AI on HR practice (non-exhaustive)
Strategic HR	• More resources devoted to strategic HR activities due to freed-up resources
Recruitment and Selection Practices	• Automate steps in the recruiting process, including posting jobs, screening candidates, and scheduling interviews • Automate and improve candidate assessment • Reduce potential recruiter bias and access a larger and more diverse candidate pool • Cost-effectiveness, reduced process time, increased agility
HR-Controlling	• Use in workforce planning • Productivity tracking
Training, Development, and Talent Management	• Offer advanced training to support transition to increased AI use • Cost-effectiveness with potential headcount reductions and increased agility • Higher level of personalisation • Potential deskilling through increased automation and reduction of skill development opportunities
Employee Relations	• Use of chatbots for recurring employee inquiries • Potential to execute some interpersonal tasks, i.e., submitting complaints • AI as decision-support system, i.e., for disciplinary actions

results. The absence or heterogeneity of definitions of AI across research reflects the often subpar methodological or overall quality of the current state of research on AI in HR. Some articles connect digital technologies with HR on a very general basis, so that no insights on the impact of AI on particular functions can be derived (e.g., Nankervis & Cameron, 2023). Others show a lack of sufficient transparency regarding database and analysis to justify conclusions (e.g., Rodney et al., 2019). This can create challenges in making inferences and establishing clear implications. In future research and when translating the findings to practice, it is worth examining precisely which type of AI is involved and what impacts may be expected from it.

Furthermore, most research on the impact of AI on HR functions and practices makes little or no use of theoretical frameworks. However, theory can help to better explain the mechanisms of relationships between constructs and variables as well as their respective contexts (Sutton & Staw, 1995). For example, we propose applying the resource-based view (Barney, 1991) to study AI as a strategic resource for HR. Institutional theory (North, 1990) can be applied to examine the influence of institutional norms on country-level implementation of AI. Finally, the job demands-resources model (Demerouti et al., 2001) is well suited for exploring how AI-driven changes in resources and demands affect individuals and the organization.

5 Practical Implications

For a successful implementation, however, Niehueser and Boak (2020, p. 128), for example, emphasise that "the design of a robust system with a simple and intuitive interface; careful planning of a pilot project with a major client; involving the team leader of the pilot from an early stage in the process; enrolling individuals to the team to implement the pilot who were seen to be flexible and open to innovation; and providing suitable training for the pilot team" is needed. In particular, the last two aspects of enrolling qualified employees and providing training fall within the expertise of HR.

Defining universally valid break-even points or benefit-risk-assessments to evaluate the introduction of AI for the average organization, would be ill-advised due to the limited and heterogeneous contexts in which applications have been tested. Anecdotal evidence is provided by Niehueser and Boak (2020), for example, who found AI beneficial for selection in a company with over 50,000 applications per day, while Malik et al. (2022) found successful implications in organizations that created the necessary software themselves in a highly customized and cost-efficient setting. When considering the introduction of AI, the organization-specific cost-benefit ratio must be assessed, despite the predominantly positive empirical findings.

HR is also an important partner for selecting key users and providing adequate training. The majority of the examined AI usage takes place in recruiting, likely due to the ease of access to and use of personal data within the boundaries of the organization. For countries with strong data protection regulation or strong internal corporate co-determination bodies, the introduction of AI in other HR areas is more complex, as it requires greater coordination and compliance. Nonetheless, the findings demonstrate that there are existing use cases for AI in HR. Practitioners should therefore contemplate the targeted use of AI in HR following a thorough analysis of company-specific enablers and barriers, as well as expected benefits. Nevertheless, research shows that there are use cases for AI in HR. Therefore, practitioners should weigh the effort of the implementation against the expected benefits and consider the targeted use of AI in HR after conducting a thorough analysis of company-specific enablers and barriers. In addition, drawing on the roles performed by the HR department as defined by Ulrich (1996), HR practitioners will need to assume a key role in facilitating the adoption of AI across the organization in their role as a strategic partner and change agent, championing and steering the organizational transformation.

6 Limitations

The purpose of this article was to provide an overview of the uses of AI in HR and its effects. To this end, we analyzed literature that describes empirically measured effects of AI on HR. This focus imposes several limitations: Firstly, the research findings are not substantiated by independent empirical evidence. Secondly, articles that do not measure any empirical effects of AI were excluded. Thirdly, it must be taken into account that there is a considerable time lag between the implementation of technical innovations and their reflection in organizational science. Therefore, we cannot claim to fully reflect the status quo. This is relevant, for example, because as generative AI has spread rapidly in recent months. This has not yet been discussed in the scientific literature.

7 Outlook

While research to date has mostly reported on efficiency gains, more far-reaching changes are possible in the future. Based on the effects of AI on HR practices reported above, AI may affect the roles of HR as proposed by Ulrich (1996). This would change the overall character of HR departments. Our findings suggest that administrative tasks could be automated easily, with AI taking on the role of administrative experts while humans function as gatekeepers or inspectors. AI so far functions as an enhanced data supplier and analyst. Although it has the potential to supplement or replace human work in HR, new practices emerge for HR professionals, such as supervising AI or deeper involvement in strategic decision making. The improved ability of AI to process data is also intertwined with its role as a change enabler, providing information for decision making, constructing employee development plans and personalized materials in record time that humans cannot offer. However, up to this point, it seems that the final decisions on the implementation of data-based recommendations concerning training or changes remain in the hands of humans. The same is true for the role of the employee champion, who speaks up and ensures that the employees' voice is heard and respected at work. While the collection and reporting of employee struggles and desires may be automated in future, the detection of mental health issues, for example, still requires the willingness of employees to share very personal information. Human empathy can be used in day-to-day interactions to collect such information faster and more efficiently. Additionally, supervisors and leaders may require a certain charisma and intuition to build trust among the workforce as well as leadership abilities to represent employee needs and concerns in the upper echelons. Those functions will presumably continue to be performed by humans in the future.

In short:

- With the introduction of AI in the HR department, a change in the roles of the HR department is likely. A shift towards a more prominent role in guiding organizational changes and a departure from administrative responsibilities is expected.
- At present, the implementation of AI is concentrated on HR practices with a sufficiently large database, such as recruiting. As HR processes are increasingly digitized, the use of AI is also anticipated to attract greater interest across other HR practices.

- When considering the introduction of AI, the organization-specific cost-benefit ratio must be assessed despite predominantly positive empirical findings. The implementation scope will also be driven by data privacy regulations in different countries, which will shape how HR data at the disposal of organizations can be leveraged to enable AI-powered HR practices.

References

Abbas, S. M., Liu, Z., & Khushnood, M. (2023). When human meets technology: Unlocking hybrid intelligence role in breakthrough innovation engagement via self-extension and social intelligence. *Journal of Computer Information Systems, 63*(5), 1183–1200. https://doi.org/10.1080/08874417.2022.2139781

Altemeyer, B. (2019). Making the business case for AI in HR: Two case studies. *Strategic HR Review, 18*. https://doi.org/10.1108/SHR-12-2018-0101

Ardichvili, A. (2022). The impact of artificial intelligence on expertise development: Implications for HRD. *Advances in Developing Human Resources, 24*(2), 78–98. https://doi.org/10.1177/15234223221077304

Barney, J. (1991). Firm resources and sustained competitive advantage. *Journal of Management, 17*(1), 99–120. https://doi.org/10.1177/014920639101700108

Bartosiak, M. L., & Modlinski, A. (2022). Fired by an algorithm? Exploration of conformism with biased intelligent decision support systems in the context of workplace discipline. *Career Development International, 27*(6/7), 601–615. https://doi.org/10.1108/CDI-06-2022-0170

Basu, S., Majumdar, B., Mukherjee, K., Munjal, S., & Palaksha, C. (2023). Artificial intelligence–HRM interactions and outcomes: A systematic review and causal configurational explanation. *Human Resource Management Review, 33*(1), 100893. https://doi.org/10.1016/j.hrmr.2022.100893

Burrell, J. (2016). How the machine 'thinks': Understanding opacity in machine learning algorithms. *Big Data & Society, 3*(1), 2053951715622512. https://doi.org/10.1177/2053951715622512

Cappelli, P., & Rogovsky, N. (2023). *Artificial intelligence in human resource management: A challenge for the human-centred agenda?* (International Labour Organization. Research Department, Ed.). ILO. https://doi.org/10.54394/OHVV4382

Cappelli, P., Tambe, P., & Yakubovich, V. (2020). Can data science change human resources? In J. Canals (Ed.), *The future of management in an AI World. Redefining purpose and strategy in the Fourth Industrial Revolution* (pp. 93–115). Springer International Publishing AG (IESE Business Collection). https://doi.org/10.1007/978-3-030-20680-2_5

Chevalier, F. (2023). *AI in HR: How is it really used and what are the risks?* https://www.hec.edu/en/knowledge/articles/ai-hr-how-it-really-used-and-what-are-risks

De Obesso, A., Pérez Rivero, C. A., & Carrero Márquez, O. (2023). Artificial intelligence to manage workplace bullying. *Journal of Business Research, 160*(C), S0148296323001716. https://doi.org/10.1016/j.jbusres.2023.113813

Del Giudice, M., Scuotto, V., Orlando, B., & Mustilli, M. (2023). Toward the human – centered approach. A revised model of individual acceptance of AI. *Human Resource Management Review, 33*(1), 100856. https://doi.org/10.1016/j.hrmr.2021.100856

Demerouti, E., Bakker, A. B., Nachreiner, F., & Schaufeli, W. B. (2001). The job demands-resources model of burnout. *Journal of Applied Psychology, 86*(3), 499–512. https://doi.org/10.1037/0021-9010.86.3.499

Hofeditz, L., Clausen, S., Rieß, A., Mirbabaie, M., & Stieglitz, S. (2022). Applying XAI to an AI-based system for candidate management to mitigate bias and discrimination in hiring. *Electronic Markets, 32*(4), 2207–2233. https://doi.org/10.1007/s12525-022-00600-9

Holton, R., & Boyd, R. (2021). 'Where are the people? What are they doing? Why are they doing it?'(Mindell) Situating artificial intelligence within a socio-technical framework. *Journal of Sociology, 57*(2), 179–195. https://doi.org/10.1177/1440783319873046

Huselid, M. A. (1995). The impact of human resource management practices on turnover, productivity, and corporate financial performance. *Academy of Management Journal, 38*(3), 635–672. https://doi.org/10.5465/256741

Jatoba, M., Ferreira, J. J., Fernandes, P., & Teixeira, J. (2023). Intelligent human resources for the adoption of artificial intelligence: A systematic literature review. *Journal of Organizational Change Management, 36*. https://doi.org/10.1108/JOCM-03-2022-0075

Kshetri, N. (2020). Artificial intelligence in developing countries. *IT Professional, 22*, 63–68. https://doi.org/10.1109/MITP.2019.2951851

Kumar, V., Rajan, B., Venkatesan, R., & Lecinski, J. (2019). Understanding the role of artificial intelligence in personalized engagement marketing. *California Management Review, 61*(4), 135–155. https://doi.org/10.1177/0008125619859317

Li, L., Lassiter, T., Oh, J., & Lee, M. K. (2021). Algorithmic hiring in practice: Recruiter and HR Professional's perspectives on AI use in hiring. https://doi.org/10.1145/3461702.3462531

Malik, A., Budhwar, P., Patel, C., & Srikanth, N. R. (2022). May the bots be with you! Delivering HR cost-effectiveness and individualised employee experiences in an MNE. *The International Journal of Human Resource Management, 33*(6), 1148–1178. https://doi.org/10.1080/09585192.2020.1859582

McKinsey & Company. (2023). *Generative AI and the future of HR*. https://www.mckinsey.com/capabilities/people-and-organizational-performance/our-insights/generative-ai-and-the-future-of-hr

Moser, C., den Hond, F., & Lindebaum, D. (2022). Morality in the age of artificially intelligent algorithms. *Academy of Management Learning & Education, 21*(1), 139–155. https://doi.org/10.5465/amle.2020.0287

Nankervis, A., & Cameron, R. (2023). Capabilities and competencies for digitised human resource management: Perspectives from Australian HR professionals. *Asia Pacific Journal of Human Resources, 61*. https://doi.org/10.1111/1744-7941.12354

Niehueser, W., & Boak, G. (2020). Introducing artificial intelligence into a human resources function. *Industrial and Commercial Training, 52*(2), 121–130. https://doi.org/10.1108/ICT-10-2019-0097

North, D. (1990). *Institutions, institutional change and economic performance (political economy of institutions and decisions)*. Cambridge University Press.

Omoigui, N. (2022). *Interserve handed £4 million fine after staff data breach*. https://www.hrmagazine.co.uk/content/news/interserve-handed-4-million-fine-after-staff-data-breach

Oracle Corporation. (2019). The 2019 State of artificial intelligence in talent acquisition. https://www.oracle.com/a/ocom/docs/artificial-intelligence-in-talent-acquisition.pdf

Oracle Corporation. (2022). *AI in human resources: The time is now*. https://www.oracle.com/a/ocom/docs/applications/hcm/oracle-ai-in-hr-wp.pdf

Poba-Nzaou, P., Galani, M., Uwizeyemungu, S., & Ceric, A. (2021). The impacts of artificial intelligence (AI) on jobs: An industry perspective. *Strategic HR Review, 20*(2), 60–65. https://doi.org/10.1108/SHR-01-2021-0003

Raisch, S., & Krakowski, S. (2021). Artificial intelligence and management: The Automation–Augmentation Paradox. *Academy of Management Review, 46*(1), 192–210. https://doi.org/10.5465/amr.2018.0072

Rodney, H., Valaskova, K., & Durana, P. (2019). The artificial intelligence recruitment process: How technological advancements have reshaped job application and selection practices. *Psychosociological Issues in Human Resource Management, 7*(1), 42–47.

Shook, E., Sage-Gavin, E., & Cantrell, S. (2019). *How companies can use employee data responsibly*. https://hbr.org/2019/02/how-companies-can-use-employee-data-responsibly

Sithambaram, R. A., & Tajudeen, F. P. (2023). Impact of artificial intelligence in human resource management: A qualitative study in the Malaysian context. *Asia Pacific Journal of Human Resources, 61*(4), 821–844. https://doi.org/10.1111/1744-7941.12356

Sutton, R. I., & Staw, B. M. (1995). What theory is not. *Administrative Science Quarterly, 40*(3), 371–384. https://doi.org/10.2307/2393788

Ulrich, D. (1986). Organizational capability as a competitive advantage: Human resource professionals as strategic partners. *Human Resource Planning, 10*(4), 169–184.

Ulrich, D. (1996). *Human resource champions: The next agenda for adding value and delivering results*. Harvard Business Press.

Waldkirch, M., Bucher, E., Schou, P. K., & Grünwald, E. (2021). Controlled by the algorithm coached by the crowd – how HRM activities take shape on digital work platforms in the gig economy. *The International Journal of Human Resource Management, 32*(12), 2643–2682. https://doi.org/10.1080/09585192.2021.1914129

Wirges, F., & Neyer, A. K. (2023). Towards a process-oriented understanding of HR analytics: implementation and application. *Review of Managerial Science, 17*(6), 2077–2108. https://doi.org/10.1007/s11846-022-00574-0

The Impact of Artificial Intelligence on Organizations and Managers: The Skills Needed for an Effective Leadership

Christian Di Prima, Simone Bevilacqua, Stefano Bresciani, and Alberto Ferraris

Abstract The chapter explores the impact of Artificial Intelligence (AI) adoption on organizations by investigating how managers have to adapt and reshape their leadership style. First, we describe how AI changes business management by shedding light on the adoption process and on AI's main impacts. These mainly relates to organizational strategy, required skills, organizational culture, and human-AI interaction. Later, we point out the main practical applications of AI within organizations and how it can improve business processes. Finally, we explore the impact of AI on leadership to understand how managers should change to lead their organizations effectively. In particular, we focus on the heterogeneous skills, factors, and characteristics that managers need to make the most of the potentialities of AI.

C. Di Prima (✉)
Department of Management, Università degli Studi di Torino, Turin, Italy

Faculty of Social Sciences and Solvay Business School, Vrije Universiteit Brussels, Brussels, Belgium
e-mail: christian.diprima@unito.it

S. Bevilacqua · S. Bresciani
Department of Management, Università degli Studi di Torino, Turin, Italy
e-mail: simone.bevilacqua@unito.it; stefano.bresciani@unito.it

A. Ferraris
Department of Management, Università degli Studi di Torino, Turin, Italy

Gnosis: Mediterranean Institute for Management Science, School of Business, University of Nicosia, Nicosia, Cyprus
e-mail: alberto.ferraris@unito.it

M. T. Del Val Núñez et al. (eds.), *Artificial Intelligence and Business Transformation*, Contributions to Management Science,
https://doi.org/10.1007/978-3-031-58704-7_10

1 Artificial Intelligence: New Era, New Businesses

Digitalization has encouraged transformations in every field, forcing organizations to evolve into digital workplaces (Cortellazzo et al., 2019). The vast amount of data, presented in several formats and generated with unprecedented speed, necessitated the evolution of new technological tools (Bresciani et al., 2021a). This process has significantly accelerated technological progress, including expanding computing capabilities and emerging new Artificial Intelligence (AI) techniques (Borges et al., 2021). According to a recent McKinsey article entitled, "What every CEO should know about generative AI", AI is defined as the ability of software to perform tasks that traditionally require human intelligence (Chui et al., 2023). Successful application of AI requires strategic planning and cooperation among various stakeholders, with business leaders playing a crucial role in ensuring the success of such initiatives. Thus, leaders must proactively allocate appropriate resources, attract and cultivate the necessary talent, collaborate with other business entities, and carefully consider the ethical implications of using AI (de Araujo et al., 2021). However, although leadership research has a history of more than a century, studies in the field of AI, specifically in the intersection of leadership and AI, is a relatively recent area of investigation. For example, the most recent analysis of leadership research published by *Leadership Quarterly* (LQ) reveals that the topic of e-leadership is present in only 0.2 percent of LQ publications between 2010 and 2019. This finding supports the idea that the connection between AI and leadership is an emerging field of research, but it is experiencing significant development and expansion (Gardner et al., 2020).

Despite understandable mistrust and initial skepticism, the relationship between humans and AI is developing in a way that may be reminiscent of what occurred with the spread of computers, starting in the 1970s, or the Internet in the following decades. As in these two previous cases, there was initially a curiosity about AI on the part of managers and entrepreneurs, followed by several initiatives that could be described as pioneering. These should be credited with first demonstrating AI's practical potential, prompting more and more companies to attempt to imitate them. According to recent surveys conducted by Forbes, entitled "The impact of Artificial Intelligence on leadership: how to leverage AI to improve decision making", currently 35% of companies have already integrated AI within their operations, while 42% of companies are considering adopting it soon (Dennison, 2023).

However, it is only recently that technological advances have meant that AI no longer requires hard-to-find resources for most organizations, financially and in terms of the necessary skills, making this technology more accessible. This has expanded the pool of users capable of approaching AI, giving a significant boost to companies operating in this sector; in fact, this market is expected to reach a value of USD 500 billion by 2024 (Weitzman, 2022). Furthermore, some analysts estimate that 50% of business decisions will be made with AI in the next five years (Mittal et al., 2022).

In any case, AI will not, at least in the short term, come to replace humans in managerial positions. On the contrary, it will be a supporting tool capable of processing and analyzing large amounts of data infinitely faster than a human being could ever do, predicting the different scenarios that might occur and the possible consequences of decisions made by the company (Kolbjørnsrud et al., 2016). However, it will always be the human counterpart that will have to assess the soundness of these predictions and make the final decision accordingly. Innovation of this magnitude will require new leadership. The company's leaders must develop new skills and evolve towards a new understanding of their role (Bresciani et al., 2021a). Conversely, rejecting technological innovations, simply not understanding them, or deciding to postpone their adoption can lead to disastrous consequences for businesses. Figure 1 below presents the framework for applying AI in business contexts and its impact on leadership.

1.1 A New Way of Running Businesses

Despite the undeniable potential of this innovation, one should not make the mistake of thinking that it is possible to integrate AI into businesses effortlessly or consider it the panacea for all ills. In fact, according to a recent survey published in the *MIT Sloan Management Review*, a lack of leadership support for AI-related initiatives was ranked as one of the main obstacles to AI adoption (Ransbotham et al., 2018). Thus, leaders must fully understand the impact of AI and how its integration into business processes will change the the company's management (Mikalef & Gupta, 2021). The implementation of AI in business operations will therefore require careful planning and strong leadership, considering several preconditions, such as (Weitzman, 2022):

- set realistic expectations and measure the possible outcomes of this innovation;
- remember that AI is not (yet) able to solve every situation or perform every task. Consequently, it will be necessary to carefully assess which specific uses are most suitable according to the characteristics of the business;
- determine whether and how AI will improve profits, productivity, efficiency, and consumer experience or reduce costs.

Since these steps may take time but are relevant in the long run, the impact of AI will need to be carefully evaluated. Business leaders must prepare their employees using actual change management methodologies (Bresciani et al., 2021a). Indeed, AI will not be able to start doing its job independently but will need the entire organization's support to develop an implementation plan consistent with the company's goals and strategy (Shrestha et al., 2019). Preparing an action plan to help employees understand how this new technology will support them will also be necessary. From the perspective of managers and employees, the consequences of implementing AI in the business reality will produce inherent changes in four main categories: the strategic transformation process, the skills and competencies

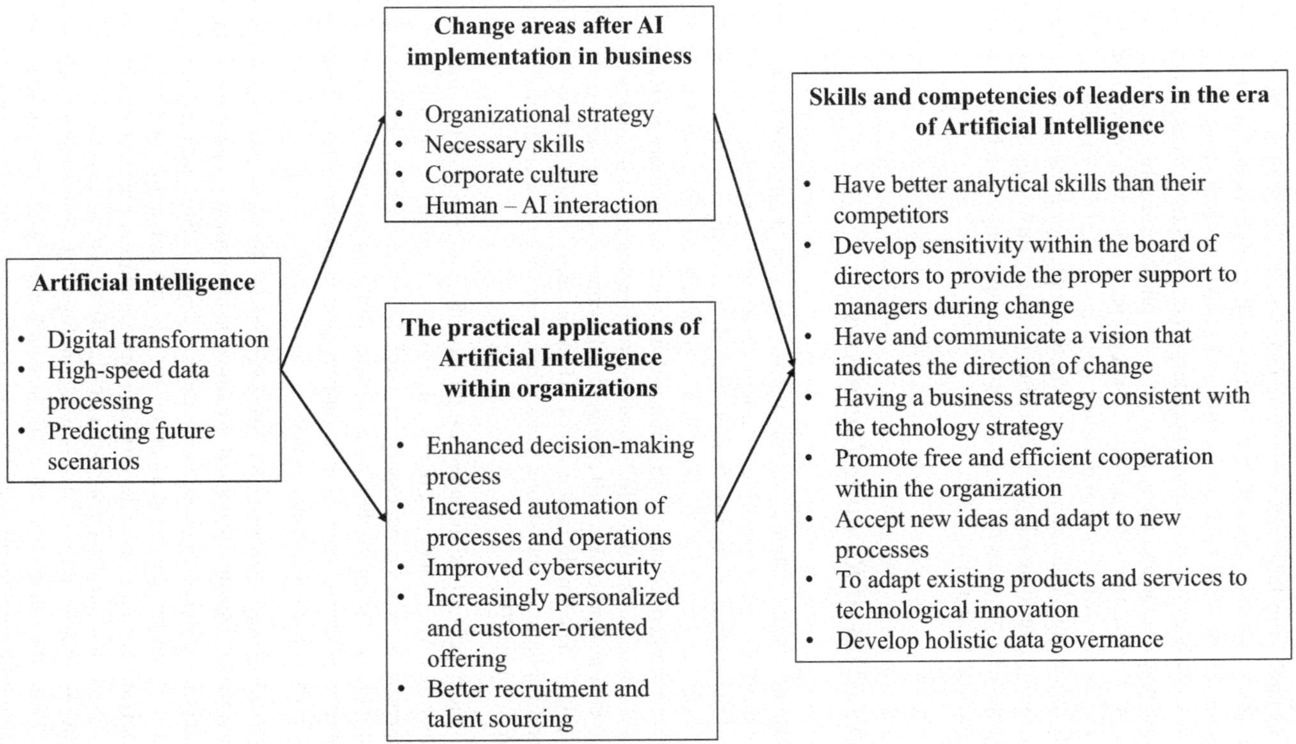

Fig. 1 AI applications in business and impact on leadership. Source: *Author's elaboration*

required, the organizational culture, and the interaction between humans and AI (Peifer et al., 2022).

1.1.1 Organizational strategy

To capitalize on the benefits offered by AI, organizations will need to conduct a careful strategic analysis aimed at the proper implementation of AI (Keding, 2021). In particular, the design of a process to manage the change resulting from the introduction of AI will be crucial and must be conceived as a long-term project. Such a strategy requires establishing detailed plans and identifying any necessary organizational changes (Mikalef & Gupta, 2021).

These transformations could relate to the company's internal structure, department collaboration, and data management. Therefore, it is essential to carefully assess the alignment of AI with the company's strategic goals before proceeding with implementation (Enholm et al., 2022). Business leaders must foster a shared understanding of AI, defining its goals according to the corporate vision. At this stage, it will also be critical for leaders to effectively communicate how the change will take place, so that the process is perceived as transparent and allows for the participation of all relevant stakeholders in its implementation, including employees and, where appropriate, union representatives. Involving these stakeholders will make addressing their doubts and concerns easier and help them understand their goals and how they should participate in the change process (Peifer et al., 2022).

1.1.2 Necessary Skills

New skills and competencies will have to be developed to take full advantage of the potential of Artificial Intelligence. First and foremost, methodological skills will be crucial. Leaders must be able to design change processes and manage their complexity (Borges et al., 2021). In parallel, it will be equally critical to emphasize the development of personal and social skills, considering the role of leaders in orchestrating the dynamics among employees and between them and AI (Enholm et al., 2022). From the initial planning phases, it will be necessary to anticipate the individual characteristics of both sides to define the future relationship between humans and technology, fostering an understanding of the role of employees within these new processes. With AI capable of drawing conclusions and making decisions autonomously, a reassignment of tasks between AI and leaders is expected. The latter will not need in-depth technical knowledge of the technological basis of AI; rather, it will be crucial to understand how to interact effectively with it. Otherwise, skills in data quality assessment and optimal use of data will become essential for managers (Smith & Green, 2018). In addition, it will be imperative for managers to understand the entire decision cycle and carefully weigh the risks associated with their choices, which are expected to increase in proportion to the growing interconnections between various processes.

1.1.3 Corporate Culture

Corporate culture is crucial in influencing the readiness for AI adoption, significantly affecting employee involvement in processes and acceptance of innovations (Pumplun et al., 2019). Innovation-oriented cultures are characterized by an openness to new ideas, showing passion and willingness to adopt innovative opportunities (Mikalef & Gupta, 2021). The presence of employees who are inclined toward learning and innovation promotes the diffusion and use of AI-related applications. This is because individuals with an innovation-oriented mindset are more likely to accept new technologies and identify new employment opportunities for AI. During the implementation phases of AI, fostering a culture that tolerates failure is essential, considering the possibility of mistakes or difficulties (Lee et al., 2019). For this purpose, leaders must play a crucial role in preparation and communication, explaining the two possible approaches of AI: as a tool to support leaders in data-driven decisions or as a substitute, delegating some decisions entirely to AI. Indeed, it is believed that organizations led by leaders who promote a culture open to innovation are more likely to integrate AI into their modus operandi (Peifer et al., 2022).

1.1.4 Human-AI Interaction

As mentioned above, AI systems can perform tasks similarly to human cognition or automate activities previously handled manually (Zheng et al., 2017). This change can affect employees' roles within the organization, requiring redefinition and the emergence of new responsibilities. Employees need to understand the purpose of AI, how it will change roles and responsibilities within the company, and whether they will need to collaborate or base their decisions on AI systems (Makarius et al., 2020). Building trust between humans and machines is a complex process, and a lack of confidence can limit AI adoption, as employees may resist change. Leaders must take preliminary actions to facilitate effective interaction between employees and AI, as understanding the technology influences trust in the AI system (Keding, 2021). These actions include deciding which tasks to entrust to humans and which to AI, considering the strengths and weaknesses of both. Although AI can improve business competitiveness, the importance of employees' social dynamics, which are crucial for innovation, should not be overlooked. Requirements such as individual protection, reliability, communication, cooperation, and social inclusion are essential for effective interaction. Managers and employees must work together to design processes that meet these requirements (Lee et al., 2019).

2 The Practical Application of Artificial Intelligence Within Organizations

The impact of AI on businesses is of considerable relevance, as it is concretely reshaping how businesses operate and opening new perspectives for development (Perifanis & Kitsios, 2023). Thanks to its ability to analyze large amounts of data, AI has the potential to improve key indicators such as revenue, efficiency, business development, digital transformation, and productivity (Keding, 2021). Its applications cover several areas, such as marketing, production management, business administration, and customer service (Bharadiya, 2023b; Wamba-Taguimdje et al., 2020). Such applications can be integrated across the entire business value chain, potentially significantly transforming multiple salient aspects of daily life (Bharadiya, 2023a). After a review of the academic literature, practical applications of AI within the organization can be identified: enhanced decision-making process, increased automation of processes and operations, improved cybersecurity, increasingly personalized and customer-oriented offering, and better recruitment and talent sourcing (Enholm et al., 2022; Kumar et al., 2019; Mikalef & Gupta, 2021; Saura et al., 2022; Weitzman, 2022).

2.1 *Enhanced Decision-Making Processes*

AI-based technologies can foster faster and better decision-making by rapidly examining vast amounts of data, identifying patterns, and providing valuable insights, predictions, and recommendations based on the data (Makarius et al., 2020). In addition, unlike individuals, AI algorithms are not subject to information overload or fatigue, enabling continuous and consistent data processing. This facilitates immediate access to the most relevant information for employees and businesses (Kumar et al., 2023). Moreover, this improved decision-making ability enables leaders to make more informed choices in significantly less time than required without AI support (Keding, 2021). Organizations that can harness the information potential of AI are in an advantageous position to grasp and react quickly to market dynamics. This reactive ability is called organizational agility and includes recognized challenges, making informed decisions, and responding promptly (Enholm et al., 2022).

2.2 *Increased Automation of Processes and Operations*

AI-based systems are distinguished by their ability to automate repetitive and standardized decisions, allowing people to focus on more strategic and creative tasks, improving their time management (Pereira et al., 2023). In addition, machines can perform tasks faster and more precisely than humans, bringing significant productivity gains in organizations, especially within manufacturing industries and supply chain management activities (Coombs et al., 2020). The replacement of human work by machines also includes reducing or eliminating errors made by employees, helping to increase transparency in operations.

2.3 Improved Cybersecurity

There is a vast field of interconnections between cyber security and AI. AI technologies, such as deep learning, can be integrated into cyber security to create intelligent models that support cyber threat classification, intrusion detection, and threat analysis (Pallathadka et al., 2023; Zhang et al., 2022). At the same time, AI models face various cyber threats that can compromise their sampling, learning, and decision-making process (Li, 2018). Consequently, business leaders need to emphasize the importance of this innovation and balance AI capabilities with human skills to build robust, adaptive, and ethical defense systems (Peifer et al., 2022).

2.4 Increasingly Personalized and Customer-Oriented Offering

AI is critical in enhancing customer satisfaction by enabling deeper understanding, more effective engagement, and a more personalized and predictive service delivery (Bresciani et al., 2021b; Kumar et al., 2019). Through AI, businesses can gain a deeper understanding of their customers' behaviors and use this increased understanding to prevent negative experiences proactively. However, the use of AI can also generate dissatisfaction among customers. For example, customers interacting with AI-based chatbots may perceive the experience as frustrating and ineffective (Castillo et al., 2021). Therefore, business leaders are responsible for integrating AI to improve the customer experience by setting strategic direction, allocating resources, driving the development of digital marketing solutions, and creating a corporate culture that fosters AI-based innovation (Saura et al., 2021).

2.5 Better Recruitment and Talent Sourcing

The traditional recruitment process in many organizations involves publishing job advertisements, analyzing resumes, and conducting interviews. However, evolving methods employ different procedures. Despite the inherent importance of human interaction in discerning the best candidates, many companies adopt recruitment and talent-sourcing solutions that use AI to identify qualified individuals efficiently. AI-based talent-sourcing solutions can interpret job descriptions in natural language and suggest the best candidates based on the qualifications indicated. Thanks to AI, these automated tools pose questions to candidates regarding their skills and experience, allowing hiring managers to receive answers from a large pool of candidates with equal effort. This approach enables companies to identify the most suitable candidates more quickly, while applicants are promptly informed of the selection process's outcome, avoiding long waits (Weitzman, 2022).

3 The Impact of Artificial Intelligence on Leadership

Successful implementation of AI requires a careful strategic plan and synergistic involvement of various stakeholders, where business leaders play a central role in determining the success of such initiatives. Leaders and managers must proactively allocate financial resources, develop new skills, collaborate with other organizational entities, and build trust. Indeed, the latter is a fundamental value in organizations during technological change, as humans trust people more than machines or, in general, processes they do not fully understand (Enholm et al., 2022). Consequently, any discussion of AI in organizations requires a profound reflection on leadership. This must be rethought since decisions based solely on intuition and previously acquired skills will be considered less reliable. Although many traits necessary for leadership will not be replicable by AI in the short term, changes in the competitive environment undeniably reshape the behaviors and skills underlying effective leadership (Peifer et al., 2022). While the "harder" elements of leadership, such as specialized skills, decision-making ability, authority, and the capacity to tackle tasks with a short-term focus, will increasingly be the preserve of AI, the "softer" elements, such as humility, adaptability, and the ability to create vision and engagement will be crucial in the development of leadership, which, with the support of AI, can become even more agile (Cortellazzo et al., 2019).

A recent Forbes article, titled "The impact of Artificial Intelligence on leadership: how to leverage Artificial Intelligence to improve decision making", points out that leadership will have to focus more on people-centred activities (Dennison, 2023). AI is, for now, able to simulate mechanical reasoning very well, but not emotional reasoning, which requires the ability to recognize, respond to, and mimic human emotions. Indeed, AI has demonstrated excellent data processing, analysis, and decision-making capabilities. However, it can still not simulate certain typical human personality traits, such as emotional stability, extroversion, and integrity, which are equally crucial for establishing effective leadership (Mikalef & Gupta, 2021). As a result, leaders will need to shift their focus from developing cognitive skills to emotional, empathic, and interpersonal relationship skills. Therefore, the leaders of this new era will have a daunting task ahead of them: they will have to make the most of technological innovation and the opportunities offered by AI without neglecting the more emotional aspects (Smith & Green, 2018). To do all this, new leadership will have to be developed: the "business" skills, that is, the ability to understand one's company, sector, and business environment, will have to be complemented by “smart” skills (Peifer et al., 2022). “Smart” leaders must thoroughly understand the mechanisms that determine thinking, commonly understood as rational, and be aware of the heuristics and biases that can condition it. Therefore, this new leadership paradigm will improve the companies' decision-making by inspiring people to increase acumen, shrewdness, and innovativeness. The next step must be to see the enterprise holistically, broadening the perspective from the mere pursuit of profitability to the quest for purpose (Kolbjørnsrud et al., 2016).

Leaders must create and capture economic value and build more sustainable and legitimate organizations, acting reasonably and responsibly. This will not have to be interpreted exclusively in the sense of “shareholder primacy”, i.e., prioritizing shareholders by ensuring their short-term profitability, but as an obligation to place the utmost loyalty and care towards the organization, ensuring its long-term sustainable value. Value that AI will be able to enhance, weaken, or even destroy, depending on how it is used (Keding, 2021).

To maximize AI’s potential, leaders (and their organizations) must acquire and consolidate a broad spectrum of heterogeneous skills, factors, and character traits. This diverse range of capabilities should include advanced technical skills in AI, a sound ethical understanding, inclusive leadership skills, an ability to interpret data, and a flair for continuous innovation (Rüth & Netzer, 2022). The main features of AI leaders are defined below:

1. Have better analytical skills than their competitors: AI must be optimally integrated into business processes and access to correct and up-to-date data from all business functions to perform at its best. To this end, companies must equip themselves with capable data analysts and provide a state-of-the-art data management structure (Peifer et al., 2022).
2. Develop sensitivity within the board of directors to provide the proper support to managers during change: as mentioned above, the new leadership should have a holistic approach, thus being able to count on the support of the entire organization. The top management will, therefore, have to make this task as easy as possible for the managers, ensuring they have the necessary support so that the whole organization is willing to work together (Enholm et al., 2022).
3. Have and communicate a vision that indicates the direction of change: integrating AI into organizations will be long and difficult. Therefore, it will be necessary to define a clear vision that always shows the direction in which the entire organization's efforts should be oriented (Peifer et al., 2022).
4. Have a business strategy consistent with the technology strategy: technological innovation must adequately support the enterprise's business soul. AI will enable business objectives to be pursued more effectively and efficiently by ensuring a holistic approach (Kaplan & Haenlein, 2019).
5. Promote free and efficient cooperation within the organization: without collaborators, even the best of leaders would be able to achieve very little. Consequently, integrating AI will necessarily have to pass through the empowerment of employees. They will have to be able to work together, creating synergies that maximize each other's talents and help them overcome any difficulties that may arise.

Since, to function at its best, AI will need access to the most significant amount of data (ideally, to the totality of corporate data), increasingly efficient and continuous collaborations must be created between different departments. Such initiatives finally overcome the old concept of "silos" that did not foresee the sharing of information, skills, and knowledge (Keding, 2021).
6. Accept new ideas and adapt to new processes: AI could even propose very "disruptive" innovations, challenging deeply rooted processes in organizations. Although an instinctive reaction to rejection is normal, managers must approach change with an elasticity of thought. It will, therefore, be necessary to reflect as much on the reasons as to why it is needed as on the positional consequences it may produce despite the foreseeable difficulties of initial adaptation (De Cremer & Kasparov, 2021).
7. To adapt existing products and services to technological innovation: AI may even suggest radical changes in the company's products or services. These might be difficult to accept, especially in the case of products that are considered successful. However, one must never forget that the fact that something works does not necessarily mean that it cannot be improved. In this sense, AI could be a great help in implementing a process of continuous improvement, which is necessary to maintain the company's competitiveness in the long run, ensuring its economic, social, and financial sustainability (Borges et al., 2021).
8. Develop holistic data governance: for AI to be able to provide valuable insights, it will ideally need data about the entire company. Therefore, creating an efficient, automated data collection structure will be necessary to reduce compilation time and the risk of human error (Alavi & Westerman, 2023).

Leaders must develop these characteristics and guide their companies by setting common goals, norms, and values. The process will be challenging and will have consequences for leaders and employees at all company levels.

4 Conclusions: Theoretical Contributions, Limitations, and Possible Future Research Directions

The concept of leadership is one of the most fascinating and debated issues when talking about business organizations. Indeed, there is an increasing need for a profound change in companies, which goes beyond the traditional theories and approaches proposed in the literature. The applications of AI have assumed a crucial role within an organization's digitization processes, thus requiring a more detailed and in-depth analysis.

The chapter aims to contribute to existing literature by defining AI applications in companies, focusing mainly on how it affects leadership. It also identifies the heterogeneous skills, factors, and attributes managers need to maximize AI's potential. By doing so, the chapter provides an in-depth compendium of previous literature regarding the relationship between AI and leadership, potentially serving as a starting point for future research on interweaving these research streams. There are significant implications for several stakeholders, especially for business leaders and practitioners in AI development. For business leaders, understanding the potential of integrating AI into the business landscape could act as an incentive to design and implement projects that employ this technology to foster resilience in business operations and improve business performance. Likewise, AI developers could gain an in-depth understanding of the synergistic effects of such technology in the business environment. As a result, AI practitioners can collaborate to devise automated and decentralized business applications that ensure optimized governance, high performance, and increased levels of security in terms of user confidentiality and privacy.

This chapter is not without limitations. A thorough literature review analyzes the leading publications on the topic. Nevertheless, possible future research could involve qualitative and quantitative analyzes, through interviews and questionnaires, to verify the application and effects of AI on employees concerning their leaders. In addition, its impact on the organizations' performance and the actual effect on business could be analyzed.

The evidence in this chapter suggests that interest in how AI impacts organizations and managers will continue to attract significant research, given its growing importance in business dynamics.

References

Alavi, M., & Westerman, G. (2023). How generative AI will transform knowledge work. *Harvard Business Review*.

Bharadiya, J. P. (2023a). A comparative study of business intelligence and artificial intelligence with big data analytics. *American Journal of Artificial Intelligence, 7*(1), 24.

Bharadiya, J. P. (2023b). Machine learning and AI in business intelligence: Trends and opportunities. *International Journal of Computer (IJC), 48*(1), 123–134.

Borges, A. F., Laurindo, F. J., Spínola, M. M., Gonçalves, R. F., & Mattos, C. A. (2021). The strategic use of artificial intelligence in the digital era: Systematic literature review and future research directions. *International Journal of Information Management, 57*, 102225.

Bresciani, S., Ciampi, F., Meli, F., & Ferraris, A. (2021b). Using big data for co-innovation processes: Mapping the field of data-driven innovation, proposing theoretical developments and providing a research agenda. *International Journal of Information Management, 60*, 102347.

Bresciani, S., Ferraris, A., Romano, M., & Santoro, G. (2021a). Prelims. In *Digital transformation management for agile organizations: A compass to sail the digital world* (pp. i–xv). Emerald Publishing Limited.

Castillo, D., Canhoto, A. I., & Said, E. (2021). The dark side of AI-powered service interactions: Exploring the process of co-destruction from the customer perspective. *The Service Industries Journal, 41*(13–14), 900–925.

Chui, M., Roberts, R., Rodchenko, T., Singla, A., Sukharevsky, A., Yee, L., & Zurkiya, D. (2023). What every CEO should know about generative AI. *Mckinsey.com*.

Coombs, C., Hislop, D., Taneva, S. K., & Barnard, S. (2020). The strategic impacts of Intelligent Automation for knowledge and service work: An interdisciplinary review. *The Journal of Strategic Information Systems, 29*(4), 101600.

Cortellazzo, L., Bruni, E., & Zampieri, R. (2019). The role of leadership in a digitalized world: A review. *Frontiers in Psychology, 10*, 1938.

de Araujo, L. M., Priadana, S., Paramarta, V., & Sunarsi, D. (2021). Digital leadership in business organizations. *International Journal of Educational Administration, Management, and Leadership*, 45–56.

De Cremer, D., & Kasparov, G. (2021). AI should augment human intelligence, not replace it. *Harvard Business Review, 18*, 1.

Dennison, K. (2023). The impact of Artificial Intelligence on leadership: How to leverage AI to improve decision-making. *Forbes.com.*

Enholm, I. M., Papagiannidis, E., Mikalef, P., & Krogstie, J. (2022). Artificial intelligence and business value: A literature review. *Information Systems Frontiers, 24*(5), 1709–1734.

Gardner, W. L., Lowe, K. B., Meuser, J. D., Noghani, F., Gullifor, D. P., & Cogliser, C. C. (2020). The leadership trilogy: A review of the third decade of the leadership quarterly. *The Leadership Quarterly, 31*(1), 101379.

Kaplan, A., & Haenlein, M. (2019). Siri, Siri, in my hand: Who's the fairest in the land? On the interpretations, illustrations, and implications of artificial intelligence. *Business Horizons, 62*(1), 15–25.

Keding, C. (2021). Understanding the interplay of artificial intelligence and strategic management: Four decades of research in review. *Management Review Quarterly, 71*(1), 91–134.

Kolbjørnsrud, V., Amico, R., & Thomas, R. J. (2016). How artificial intelligence will redefine management. *Harvard Business Review, 2*(1), 3–10.

Kumar, S., Lim, W. M., Sivarajah, U., & Kaur, J. (2023). Artificial intelligence and blockchain integration in business: Trends from a bibliometric-content analysis. *Information Systems Frontiers, 25*(2), 871–896.

Kumar, V., Rajan, B., Venkatesan, R., & Lecinski, J. (2019). Understanding the role of artificial intelligence in personalized engagement marketing. *California Management Review, 61*(4), 135–155.

Lee, J., Suh, T., Roy, D., & Baucus, M. (2019). Emerging technology and business model innovation: The case of artificial intelligence. *Journal of Open Innovation: Technology, Market, and Complexity, 5*(3), 44.

Li, J. H. (2018). Cyber security meets artificial intelligence: A survey. *Frontiers of Information Technology & Electronic Engineering, 19*(12), 1462–1474.

Makarius, E. E., Mukherjee, D., Fox, J. D., & Fox, A. K. (2020). Rising with the machines: A sociotechnical framework for bringing artificial intelligence into the organization. *Journal of Business Research, 120*, 262–273.

Mikalef, P., & Gupta, M. (2021). Artificial intelligence capability: Conceptualization, measurement calibration, and empirical study on its impact on organizational creativity and firm performance. *Information & Management, 58*(3), 103434.

Mittal, N., Irfan, S., & Beena, A. (2022). Fueling the AI transformation: Four key actions powering widespread value from AI, right now. *Deloitte's State of AI in the Enterprise, 5th Edition report, UK, retrieved in Jan 5*, 2023.

Pallathadka, H., Ramirez-Asis, E. H., Loli-Poma, T. P., Kaliyaperumal, K., Ventayen, R. J. M., & Naved, M. (2023). Applications of artificial intelligence in business management, e-commerce and finance. *Materials Today: Proceedings, 80*, 2610–2613.

Peifer, Y., Jeske, T., & Hille, S. (2022). Artificial intelligence and its impact on leaders and leadership. *Procedia Computer Science, 200*, 1024–1030.

Pereira, V., Hadjielias, E., Christofi, M., & Vrontis, D. (2023). A systematic literature review on the impact of artificial intelligence on workplace outcomes: A multi-process perspective. *Human Resource Management Review, 33*(1), 100857.

Perifanis, N. A., & Kitsios, F. (2023). Investigating the influence of artificial intelligence on business value in the digital era of strategy: A literature review. *Information, 14*(2), 85.

Pumplun, L., Tauchert, C., & Heidt, M. (2019). A new organizational chassis for artificial intelligence-exploring organizational readiness factors. In *Proceedings of the 27th European Conference on Information Systems* (ECIS), Stockholm & Uppsala, Sweden.

Ransbotham, S., Gerbert, P., Reeves, M., Kiron, D., & Spira, M. (2018). Artificial intelligence in business gets real. *MIT Sloan Management Review*.

Rüth, R., & Netzer, T. (2022). The impact of AI on leadership: New strategies for a human-machine-cooperation. *Global Journal of Management and Business Research, 22*(A8), 1–7.

Saura, J. R., Ribeiro-Soriano, D., & Palacios-Marqués, D. (2021). Setting B2B digital marketing in artificial intelligence-based CRMs: A review and directions for future research. *Industrial Marketing Management, 98*, 161–178.

Saura, J. R., Ribeiro-Soriano, D., & Palacios-Marqués, D. (2022). Assessing behavioral data science privacy issues in government artificial intelligence deployment. *Government Information Quarterly, 39*(4), 101679.

Shrestha, Y. R., Ben-Menahem, S. M., & Von Krogh, G. (2019). Organizational decision-making structures in the age of artificial intelligence. *California Management Review, 61*(4), 66–83.

Smith, A. M., & Green, M. (2018). Artificial intelligence and the role of leadership. *Journal of Leadership Studies, 12*(3), 85–87.

Wamba-Taguimdje, S. L., Fosso Wamba, S., Kala Kamdjoug, J. R., & Tchatchouang Wanko, C. E. (2020). Influence of artificial intelligence (AI) on firm performance: The business value of AI-based transformation projects. *Business Process Management Journal, 26*(7), 1893–1924.

Weitzman, T. (2022). The top five ways AI is transforming business. *Forbes.com*.

Zhang, Z., Ning, H., Shi, F., Farha, F., Xu, Y., Xu, J., Zhang, F., & Choo, K. K. R. (2022). Artificial intelligence in cyber security: Research advances, challenges, and opportunities. *Artificial Intelligence Review*, 1–25.

Zheng, N. N., Liu, Z. Y., Ren, P. J., Ma, Y. Q., Chen, S. T., Yu, S. Y., Xue, J. R., Chen, B. D., & Wang, F. Y. (2017). Hybrid-augmented intelligence: Collaboration and cognition. *Frontiers of Information Technology & Electronic Engineering, 18*(2), 153–179.

Technologies and Team Management to Increase Productivity in a Digital Age

Antonio de Lucas Ancillo, Sorin Gavrila Gavrila, José Ángel Tébar-Sáez, and Arturo Orea Rocha

Abstract The digital era has brought a paradigm shift in business operations, requiring a fundamental transformation of technologies and team management to drive productivity. In this context, organizations are exploring innovative solutions to improve efficiency, streamline processes and adapt to the changing landscape. Embracing digital technologies and adopting data-driven strategies has become paramount to remain competitive. However, technology alone cannot achieve maximum productivity. Effective team management plays a critical role in maximizing the potential of these technologies. Leadership must foster a culture of innovation, encouraging continuous learning and experimentation. Cross-functional teams can leverage diverse perspectives, enhancing problem solving and creativity. This is why the chapter addresses all of these issues, with a focus on real-time communication platforms and collaboration tools. Each of these is addressed and the challenges of implementing the tools and systems are discussed.

1 Introduction

In the current business environment, the convergence of new technologies and effective human resources management has become a fundamental pillar of organizational success. This chapter explores how cutting-edge technologies can be harnessed to optimize the productivity of work teams, fostering efficient collaboration and the achievement of strategic objectives. Actionable approaches and real-world examples illustrate how leaders and managers can adapt to this evolving digital environment and lead teams to optimal performance.

A. de Lucas Ancillo · S. G. Gavrila · J. Á. Tébar-Sáez (✉) · A. O. Rocha
University of Alcalá, Faculty of Economics, Business and Tourism, Department of Economics and Business, Madrid, Spain
e-mail: antonio.lucas@uah.es; sorin.gavrila@uah.es; angel.tebar@edu.uah.es; arturo.orea@uah.es

M. T. Del Val Núñez et al. (eds.), *Artificial Intelligence and Business Transformation*, Contributions to Management Science,
https://doi.org/10.1007/978-3-031-58704-7_11

1.1 Context of the Technological Revolution and Its Importance in Productivity

In this digital era, an unprecedented technological revolution is transforming the way of living, working and interacting (Tandon & Das, 2023). The increasing interconnectedness of devices, the explosion of data, and the automation of processes are redefining the global economy and the dynamics of both organizations *per se* and human resource management. This fascinating landscape is explored in detail, examining how new technologies are reshaping team management strategies and how professionals play a key role in this transition to a more digitized and future-oriented work environment.

The importance of effective team and technology management on productivity cannot be underestimated in the technological era. Organizations must make the most of the technological tools available to increase efficiency and stay relevant in an increasingly competitive and evolving business environment. Team management is also a critical factor, in aligning the skills and motivation of employees with the company's strategic objectives (Aarthi & Sujatha, 2022). When these two disciplines are properly combined, it creates an environment conducive to innovation, collaboration, and agile adaptation to change, ultimately leading to a significant increase in organizational productivity. This chapter briefly explores these two interconnected tools, which are essential for success in the contemporary business world, and how leaders can optimize their management of teams and technologies to drive productivity and sustainable growth.

1.2 Challenges and Solutions for Global Team Management

Global organizations, in particular, need a higher level of digitization in team management. The technological revolution and the tools it provides are key resources in solving the challenges that companies with global presence need to face. This is one of the best examples of the application of digital tools that will be discussed in the following sections. Managing global teams presents a unique set of challenges, in that one of the main obstacles is the cultural and geographic diversity of teams, which can lead to differences in communication, expectations, and work dynamics. To address this challenge, it is essential to promote cross-cultural awareness, train leaders to understand and respect cultural differences, and foster effective communication in a virtual environment.

Another important challenge is coordination and collaboration in a dispersed global environment. Here, technology also plays a crucial role in providing online collaboration tools, project management platforms, and advanced communication systems. Implementing these technological solutions must go hand in hand with creating clear policies and processes that guide collaboration and decision-making in distributed teams (Bagga et al., 2023). Ultimately, successful management of global teams requires a careful balance between technology and interpersonal skills, focusing on adaptability and flexibility to meet changing challenges in an increasingly connected world.

2 Enabling Technologies for Team Management

This section presents various tools and systems for managing teams and information flows across different organizational levels. The tools are detailed according to different aspects or levels within the organization (Fig. 1).

Recently, these tools have been used to implement artificial intelligence (AI) and data analytics that transform team management (Chowdhury et al., 2023) by providing valuable insights into performance and data-driven decision-making. AI can help identify patterns in team behavior, enabling leaders to make informed decisions about resource allocation and productivity optimization. Together, these enabling technologies not only simplify team management but also offer opportunities to improve team performance and efficiency continuously in an ever-evolving digital environment.

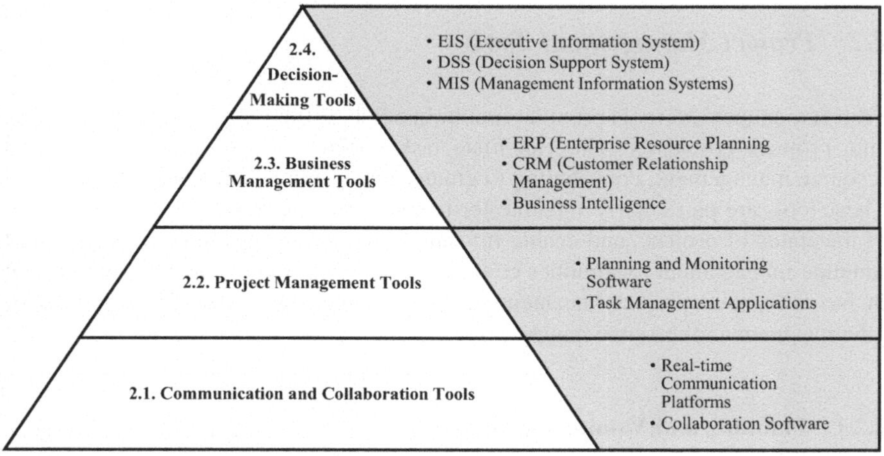

Fig. 1 Enabling technologies for team management

2.1 Real-Time Communication Platforms and Collaboration Tools

One of the most prominent tools is online collaboration platforms, which enable real-time communication, task management, and efficient collaboration across geographically distributed teams. These platforms foster transparency, accessibility, and organization, which in turn increase productivity and team cohesion.

Communication tools such as Slack, Microsoft Teams, or Zoom play an essential role in managing teams, enabling real-time communication regardless of geographical location. These tools not only improve communication efficiency but also foster a sense of community and connection among team members (Dinh et al., 2021). Similarly, cloud-based Collaboration software like Google Workspace, Microsoft 365, or Dropbox Business revolutionizes team collaboration (Khalil & Khalil, 2020), allowing simultaneous content-creation and the editing of documents, spreadsheets, and presentations from anywhere.

These tools integrate with other services like project management tools and cloud storage, centralizing information, streamlining data retrieval, and sharing. This integration enhances team responsiveness to business challenges and promotes efficient coordination in a globalized environment. Moreover, the security and data backup features of these cloud systems address growing concerns about cybersecurity and business continuity, making them indispensable in modern business operations.

2.2 Project Management Tools

Project management tools provide a centralized platform to plan, monitor, and execute projects efficiently. They facilitate task-assignment, progress tracking, and resource management, contributing to greater transparency and team collaboration. These tools are particularly valuable for team leaders, as they provide a clear view of the status of projects and enable informed decisions on resource allocation and timeline management (Majumder et al., 2022). Agility and adaptability are crucial in business, then project management tools become an essential component of effective team and resource management.

2.2.1 Planning and Monitoring Software

These applications allow organizations to design, organize, and track projects efficiently. They facilitate schedule creation, task assignment, and resource management, promoting effective coordination among team members. In addition, these solutions provide real-time visibility of the progress of projects, allowing leaders to make informed and quick decisions to adjust strategies if necessary (Kafile, 2021).

Identifying the critical path of any project is crucial to its execution, as it shows where most attention or effort must be focused in order to achieve objectives on time. In an increasingly dynamic and competitive business world, where effective management of time and resources is essential, project planning and tracking software become indispensable allies in optimizing productivity and ensuring the success of business initiatives.

2.2.2 Task Management Applications

These virtual spaces allow teams to create, assign, and track tasks and projects collaboratively. They provide a clear view of individual and collective responsibilities, which promotes organization and efficiency in task execution. Task management platforms also facilitate communication within the team by allowing real-time feedback and updates. This enhances collaboration and ensures that all members are aligned on goals and deadlines (Nundlall & Nagowah, 2021). They can also serve as a performance measurement tool. In a work environment where agility and coordination are critical, these platforms have become essential tools for managing teams and resources, boosting productivity and responsiveness to business challenges.

2.3 Business Management Tools

Business management tools, also known as ERP (Enterprize Resource Planning) and CRM (Customer Relationship Management), represent a key aspect of the technological arsenal for managing teams in the technological era (Roba & Maric, 2023). ERP systems integrate and automate key business processes, from financial and human resource management to supply chain and production. On the other hand, CRM systems focus on customer relationship management and automating the sales and marketing processes. Both categories of business management tools not only optimize internal business operations but also improve communication and collaboration between teams, resulting in greater efficiency and competitiveness.

2.3.1 ERP (Enterprize Resource Planning)

ERP systems fully integrate and automate various business processes, including accounting, inventory management, human resources, and project management. This gives organizations a complete, real-time view of their operations, facilitating informed decision-making and resource optimization (Putra et al., 2021; Skare & Soriano, 2021). In addition, ERP systems also promote efficiency by standardizing workflows and structuring the exchange of information between departments in an organization. Leaders can instantly access up-to-date, accurate data, improving agility and responsiveness to business challenges. Information is an invaluable asset, so ERP systems have become an essential tool for business management, boosting efficiency, productivity, and competitiveness.

2.3.2 CRM (Customer Relationship Management)

CRM systems are an essential internal working tool for sales teams. These technology solutions focus on customer relationship management, enabling organizations to efficiently manage information related to customers, sales, marketing, and customer service (Peddireddy & Peddireddy, 2023). Through CRM, companies can track customer interactions, facilitating personalized relationships and informed decision-making by identifying important factors such as loyalty, strategic business relationships, and growth opportunities in the market. Therefore, by contributing to the definition of business objectives, CRM favors increased efficiency among the workers involved, boosting team management.

It also improves collaboration between departments such as sales, marketing, and customer service by providing a centralized platform for sharing data and knowledge about customers. This not only increases operational efficiency but also boosts customer satisfaction and retention. In summary, CRM systems are essential tools in managing teams and resources, improving customer relationship management, and fostering effective collaboration between teams to achieve business success (Das & Hassan, 2022).

2.3.3 Business Intelligence

Business intelligence (BI) is a critical discipline in managing teams and resources, as it focuses on the collection, analysis, and presentation of business data to make informed, strategic decisions. BI tools enable organizations to transform large volumes of data into valuable information, providing detailed insight into internal processes, team performance, and market trends.

BI is essential for team management as it enables leaders to assess performance objectively, identify areas for improvement, and make data-driven decisions (Conboy et al., 2020). In addition, it supports the collection and interpretation of the information that shapes key business performance KPIs. Through these tools, reports and dashboards can be created and made accessible to the entire team, which greatly assists management in making strategic and operational decisions. In the age of technology, where information is a key asset, business intelligence becomes an essential tool for human resource management and decision-making in an ever-changing business environment.

2.3.4 Integration of Management Systems

In addition to the main business management systems mentioned above, various systems specialize in specific tasks or departments, such as supplier management or logistics and supply chain management.

Each management system has its own databases from which it is managed. Management systems need to be integrated so that the different databases are interlinked and the information is updated globally for all systems (Markov, 2023). Only in this way can the digitization of the management of all the departments converge in the main ERP management systems, with key indicators that favor decision-making, in turn making global management possible.

2.4 Decision-Making Tools

The decision-making tools, such as data analytics platforms and dashboards, along with the BI systems mentioned above, enable organizations to collect, process, and visualize relevant information effectively (Alinezhad & Khalili, 2019). This facilitates data-driven decision-making, which is essential in a highly competitive and constantly changing business environment.

It can take the form of status and trend reports, or dashboards, which should be clear, concise, and highly visual. These tools provide the ability to assess project performance, measure the impact of decisions made, and forecast trends. This provides leaders with a solid basis for evaluating strategies, optimizing resources, and responding quickly to the dynamics of the environment. In a digital world where information is abundant but the ability to make informed decisions is critical, these tools have become essential allies in managing teams and resources, driving business effectiveness and success.

2.4.1 MIS (Management Information Systems)

These systems integrate technology, processes, and human resources to collect, process, and present relevant information for decision-making in an organization. MIS ranges from basic data management to advanced reporting and analysis, enabling leaders and managers to gain a comprehensive, real-time view of the company's operational, financial, and strategic performance (Sinulingga et al., 2023).

In addition, MIS plays a vital role in team coordination and collaboration by facilitating communication and shared access to critical information. This promotes informed decision-making, resource optimization, and continuous improvement in project and process management. In an increasingly data-driven and technology-driven business environment, MIS are becoming essential tools for driving organizational efficiency, productivity, and competitive advantage.

2.4.2 DSS (Decision Support Systems)

Decision Support Systems (DSS) systems are specifically designed to assist leaders and managers in making strategic and tactical decisions. DSSs incorporate data from various sources, analyze it, and present it in a comprehensible format, which facilitates the evaluation of different scenarios and options (Phillips-Wren et al., 2021). These tools not only improve the quality of decisions but also streamline the decision-making process, which is especially valuable in a business environment characterized by rapid change and information complexity.

A decision support application might typically gather and present information on inventories of information assets (including legacy and relational data sources or data warehouses), comparative sales figures between one period and the next, and projected revenue figures based on product sales assumptions.

In addition, DSSs foster collaboration among team members and departments by providing a common platform for the analysis and discussion of data and results. This promotes a shared understanding and a data-driven approach to problem-solving and strategy formulation. In short, DSSs have become essential tools in team and resource management, enabling organizations to make informed, efficient, and strategic decisions in an increasingly complex and dynamic business environment.

2.4.3 EIS (Executive Information Systems)

Executive Information Systems (EIS) systems are designed to provide real-time strategic information that assists senior managers in making critical decisions for the organization. EISs present key data in a concise, visually appealing manner, enabling leaders to quickly understand the current state of the business and relevant trends (Parulian et al., 2023).

EISs also facilitate communication and collaboration across the organization by enabling executives to share strategic information effectively with other leaders and teams. These tools are highly customizable, allowing executives to tailor the information presented to their specific needs and areas of focus. In short, EISs have become an invaluable resource for senior executives, providing them with a complete and up-to-date view of the organization and enabling them to make critical strategic decisions.

3 Challenges Faced in Implementing the Tools and Systems

While team management technologies can provide numerous ways of increasing productivity in companies, they also present challenges that companies must address. Some of the common challenges are as follows.

a) The introduction of new technologies often meets resistance from employees accustomed to traditional work methods. It is crucial to have a well-structured change plan that includes effective communication, training, and demonstrations of the benefits the technology will bring.
b) Integrating new management technologies with existing systems can be difficult, especially if there are technological incompatibilities. The company must thoroughly assess its existing technology infrastructure and seek compatible or easily integrated solutions. Management should take into account that integrating and/or implementing systems can take between six to nine months. In addition, managing teams often involves handling sensitive and confidential information, so data security is a major concern: companies must implement robust security measures such as encryption, user authentication, and access policies.
c) The absence of the skills and knowledge necessary to use team management technologies effectively can hinder their adoption. This can be solved by providing training programs and educational resources to train employees in the proper use of tools, and foster a culture of continuous learning.
d) Every company has its own unique processes and workflows, so finding team management solutions that fit specific needs can be problematic. Flexible and customizable solutions will allow adaptation to the company's internal processes.
e) Implementing new technologies often requires a change in organizational culture, which can be difficult to achieve. The solution would be to foster a mindset open to change, to recognize and reward the successful adoption of new technologies, and to lead by example from senior management.
f) The initial investment in management technologies can be considerable, and some companies may find it difficult to justify. Therefore, a clear analysis of the anticipated return on investment (ROI), and the demonstration of how the technology will improve efficiency and productivity in the long term, are essential.

4 Summary

In the context of team and resource management, various technological tools have become key assets for business success. These tools cover a wide range of applications, from project planning and tracking systems to executive information systems, focused on optimizing communication, data-driven decision-making, and efficient coordination.

A comprehensive review of the current state of the art was conducted, identifying the main tools used by companies and organizations to enhance their productivity. An analysis of these tools was conducted, covering the different organizational levels and business requirements. The review started at the general, global level, with collaboration tools relevant to all members of the organization. Programs such as Slack, Microsoft Teams, Zoom, or Discord were identified. Another key factor at a general level is the use of collaboration software in the cloud, which allows global access to information and simultaneous work from anywhere in the world. Examples included Google Workspace, Microsoft 365, and Dropbox Business.

With regard to resources focused on executive business management, three main categories can be distinguished: ERP, CRM, and BI. ERP systems optimize the organization's management and planning of resources and as serve as financial information systems. CRM systems are more commonly used for efficient customer and supplier relationship management at the operational level. Business Intelligence tools are used and recommended widely. They enable organizations to transform large volumes of data into valuable information, providing detailed insights into internal processes, team performance, and market trends. Software products were identified as relevant to lower-level management functions such as project management, planning, monitoring, and personnel management. Management support tools that facilitate decision-making at different levels, such as EIS, DSS, and MIS, were also identified. Using EIS systems, executives can receive strategic information from which DSSs then assist with high-level decisions. Then, operational strategies flow directly down to intermediate managers who can align all their teams by using MIS systems.

The tools, platforms, systems, and software presented here enhance organizations' ability to navigate an ever-changing business environment, improving productivity, competitiveness, and agility. The strategic adoption of these tools is essential for effectively managing teams and resources in the technological revolution, providing the necessary foundation for informed decision-making and achieving business objectives.

This chapter is intended to serve as an extension and update of the existing tools in the current theoretical framework. Future studies on the subject could focus on measuring the real impact of these tools, or on how these tools are perceived by different actors in organizations. To sum up, by adopting these new solutions and addressing their challenges holistically, companies can maximize the benefits of team management technologies and significantly improve productivity.

References

Aarthi, G., & Sujatha, S. (2022). Future team leadership and empowerment in the performance of measuring virtual team productivity with information technology. *International Journal of Professional Business Review, 7*(2), 6.

Alinezhad, A., & Khalili, J. (2019). *New methods and applications in multiple attribute decision making (MADM)* (Vol. 277, pp. 103–108). Springer. https://doi.org/10.1007/978-3-030-15009-9

Bagga, S. K., Gera, S., & Haque, S. N. (2023). The mediating role of organizational culture: Transformational leadership and change management in virtual teams. *Asia Pacific Management Review, 28*(2), 120–131. https://doi.org/10.1016/j.apmrv.2022.07.003

Chowdhury, S., Dey, P., Joel-Edgar, S., Bhattacharya, S., Rodriguez-Espindola, O., Abadie, A., & Truong, L. (2023). Unlocking the value of artificial intelligence in human resource management through AI capability framework. *Human Resource Management Review, 33*(1), 100899. https://doi.org/10.1016/j.hrmr.2022.100899

Conboy, K., Mikalef, P., Dennehy, D., & Krogstie, J. (2020). Using business analytics to enhance dynamic capabilities in operations research: A case analysis and research agenda. *European Journal of Operational Research, 281*(3), 656–672. https://doi.org/10.1016/j.ejor.2019.06.051

Das, S., & Hassan, H. K. (2022). Impact of sustainable supply chain management and customer relationship management on organizational performance. *International Journal of Productivity and Performance Management, 71*(6), 2140–2160. https://doi.org/10.1108/IJPPM-08-2020-0441

Dinh, J. V., Reyes, D. L., Kayga, L., Lindgren, C., Feitosa, J., & Salas, E. (2021). Developing team trust: Leader insights for virtual settings. *Organizational Dynamics, 50*(1), 100846.

Kafile, M. (2021). Project planning and scheduling in the face of the Fourth Industrial Revolution (4IR). *Journal of Business Administration Research, 4*(3). https://doi.org/10.30564/jbar.v4i3.3398

Khalil, C., & Khalil, S. (2020). Exploring knowledge management in agile software development organizations. *International Entrepreneurship and Management Journal, 16*(2), 555–569. https://doi.org/10.1080/01930826.2020.1760569

Majumder, S., Majumder, S., & Biswas, D. (2022). Impact of effective construction planning in project performance improvement. *Quality & Quantity, 56*(4), 2253–2264. https://doi.org/10.1007/s11135-021-01224-5

Markov, D. A. (2023). Automated management systems: Problems of implementation and integration. *Technoeconomics: An International Journal, 2*(1). https://doi.org/10.57809/2023.2.1.4.5

Nundlall, C., & Nagowah, S. D. (2021). Task allocation and coordination in distributed agile software development: A systematic review. *International Journal of Information Technology, 13*, 321–330. https://doi.org/10.1007/s41870-020-00543-4

Parulian, R., Ali, H., & Sawitri, N. N. (2023). Executive support system for business and employee performance: Analysis of the ease of use of information system, user satisfaction and quality of information system. *International Journal of Psychology and Health Science, 1*(2), 65–77. https://doi.org/10.38035/ijphs.v1i2.203

Peddireddy, A., & Peddireddy, K. (2023). Next-Gen CRM sales and lead generation with AI. *International Journal of Computer Trends and Technology, 71*(3), 21–26. https://doi.org/10.14445/22312803/IJCTT-V71I3P104

Phillips-Wren, G., Daly, M., & Burstein, F. (2021). Reconciling business intelligence, analytics and decision support systems: More data, deeper insight. *Decision Support Systems, 146*, 113560. https://doi.org/10.1016/j.dss.2021.113560

Putra, D. G., Rahayu, R., & Putri, A. (2021). The influence of Enterprise Resource Planning (ERP) implementation system on company performance mediated by organizational capabilities. *Journal of Accounting and Investment, 22*(2), 221–241. https://doi.org/10.18196/jai.v22i2.10196

Roba, G. B., & Maric, P. (2023). AI in customer relationship management. In *Developments in information and knowledge management systems for business applications* (Vol. 7, pp. 469–487). Springer Nature Switzerland. https://doi.org/10.1007/978-3-031-25695-0_21

Sinulingga, M., Djati, P., Thamrin, S., Saragi, H. J. R., Riyadi, B. S., & Ubayanto, T. (2023). Antecedents and consequences of smart management information system for supervision to improve organizational performance. *International Journal of Membrane Science and Technology, 10*(2), 816–824. https://doi.org/10.15379/ijmst.v10i2.1262

Skare, M., & Soriano, D. R. (2021). How globalization is changing digital technology adoption: An international perspective. *Journal of Innovation & Knowledge, 6*(4), 222–233. https://doi.org/10.1016/j.jik.2021.04.001

Tandon, S., & Das, S. (2023). Digital revolution complements sustainable goal achievement during and after pandemic: A research agenda. *New Normal in Digital Enterprises: Strategies for Sustainable Development, 227-246*. https://doi.org/10.1007/978-981-19-8618-5_11

The Core Competencies of Future Leaders: Opportunities and Challenges of Artificial Intelligence for Business Schools

Andy Coleman and Katerina Beta

Abstract The concept of artificial intelligence has existed since the 1950s and is commonly taken to be an amalgamation of diverse technologies working in unison to enable machines to sense, comprehend, act, and learn with a level of intelligence akin to that of humans (Accenture, What is artificial intelligence (AI) & why is it important? 2022). Recent years have witnessed increasingly rapid developments in AI technologies. Consequently, AI now plays an indispensable role in driving social progress and yielding transformative outcomes in many industries, including higher education (Duan et al., Journal of Industrial Integration and Management 15:201–206, 2019). This chapter critically explores the impact AI is already having on business schools, and implications for the future. It discusses how AI is transforming day-to-day operations in business schools, including in the practical management of core operations and business processes, and the core business of education. In doing so, it examines the transformational impact of AI on every aspect of teaching and learning, including the curation of knowledge, students' interactions with resources and the ways in which learners demonstrate their mastery of the subject. The chapter also explores the specific implications AI has for leaders in the future, and what this means in terms of business schools' flagship MBA (Master of Business Administration) programmes.

1 Introduction

The concept of artificial intelligence has existed since the 1950s and is commonly taken to be an amalgamation of diverse technologies working in unison to enable machines to sense, comprehend, act, and learn with a level of intelligence akin to

A. Coleman (✉) · K. Beta
School of Strategy and Leadership, William Morris Building, Coventry University, Coventry, UK
e-mail: ac8156@coventry.ac.uk

M. T. Del Val Núñez et al. (eds.), *Artificial Intelligence and Business Transformation*, Contributions to Management Science,
https://doi.org/10.1007/978-3-031-58704-7_12

that of humans (Accenture, 2022). Recent years have witnessed increasingly rapid developments in AI technologies. Consequently, AI now plays an indispensable role in driving social progress and yielding transformative outcomes in many industries, including higher education (Duan et al., 2019).

This chapter critically explores the impact AI is already having on business schools, and implications for the future. It discusses how AI is transforming day-to-day operations in business schools, including in the practical management of core operations and business processes, and the core business of education. In doing so, it examines the transformational impact of AI on every aspect of teaching and learning, including the curation of knowledge, students' interactions with resources and the ways in which learners demonstrate their mastery of the subject. The chapter also explores the specific implications AI has for leaders in the future, and what this means in terms of business schools' flagship MBA (Master of Business Administration) programmes.

2 Methodology

The chapter utilises an integrated literature review method to explore the most important material on its central themes of the core competencies AI demands for future leaders, and the opportunities and challenges AI presents for business schools. However, in doing so it seeks to go beyond the aim of a more 'traditional' literature review to summarise the existing literature in the field, and instead adopts a "distinctive form of research that uses existing literature to create new knowledge" (Torraco, 2016, p. 405). Further originality is added by the inclusion of our own reflections, from our perspective as professional educators on MBA programmes, where we have already witnessed personally how AI is changing the focus for, and nature of learning for leaders of the future.

3 Background

AI plays an indispensable role in driving social progress and has yielded transformative outcomes in enhancing labour efficiency, reducing workforce expenditures, optimising human resource structures, and generating fresh job demands (Duan et al., 2019; Foffano et al., 2023). From an operational perspective, industries with well-established data foundations, including finance, healthcare, automotive, and retail, have developed relatively sophisticated AI applications (Zhang & Lu, 2021). In the automotive sector, autonomous driving stands as a product of the profound integration of cutting-edge information technologies from across the automotive industry, AI, and the Internet of Things. Within the financial domain, the AI market is rapidly increasing. Leveraging machine learning techniques, AI is able to predict both risks and stock market trends. Financial institutions use machine learning methodologies to manage financial risks, integrate diverse data sources, and offer real-time risk alerts to individuals (Chen et al., 2020). In the healthcare sector, AI-related algorithms provide medical support in a variety of ways, such as the effective diagnosis of a range of conditions (including the detection of cancer) and the development of new drugs to address them (Kumar et al., 2023). Within the retail industry, offline physical stores use AI to realise genuine unmanned retail operations, thereby lowering costs and significantly enhancing efficiency. For example, AmazonGo, a smart physical retail store established by e-commerce giant Amazon, has rapidly driven smart retail to the forefront (Vollero et al., 2023). In the media sphere, content communication robots and brand communication robots enable one-click generation of user-desired content, allowing for the creation of up to 10,000 articles in a single minute (Nunavath & Goodwin, 2018).

The higher education sector, like the above industries, has also been highly impacted by AI technologies. Indeed, the Educase Horizon Report (2021) identifies AI as one of six pivotal technologies and practices, set to have a profound influence on teaching and learning in higher education (others include Blended and Hybrid Course Models, Learning Analytics, Microcredentials, Open Education Resources, and Quality Online Learning (Haderer & Ciolacu, 2022)). Indeed, its impacts can already be felt; the continued evolution of digital education for instance is facilitated by the integration of AI, to create optimized learning environments (Haderer & Ciolacu, 2022), designed to enhance students' learning environment. Similarly, AI is being widely utilised to support a wide variety of management and administration functions. These themes are explored further later in this chapter.

More broadly, recent discussions on AI and higher education have highlighted two key aspects. First, AI is seen as a means to address current challenges in teaching, learning, and student success.

To date, most applications of AI in higher education appear to have focused on this area and have often utilised bot-based technology to do so. For example, at Durham University in the UK, staff have developed an AI-based 'student engagement platform' called Holly to enhance student success during the admissions process. Elsewhere, The University of British Columbia has initiated the Language Chatsim project, recognizing the effectiveness of immersive language learning.

Second, there is a growing imperative to fundamentally reimagine the curriculum and associated academic programs to better serve 'Generation AI'. Part of this is a reflection of the challenges AI itself is already presenting within the sector. Adeshola and Adepoju (2023) for example, highlight how ChatGPT has facilitated an explosion in cheating, as many students have been quick to recognise its potential as a mechanism for completing their assignments. In response, educators are increasingly exploring alternative approaches to assessing learning, for instance by adopting more authentic methods, some of which are described later in this chapter. Unsurprisingly then, opinions on AI within many universities are often mixed. As Hirsh-Pasek and Blinkoff (2023) argue, "ChatGPT is only dangerous if our educational system is pursuing rubric points and not knowledge". Additionally, Barari and Kumar (2023) argue that ChatGPT supports language translation applications due to its ability to function in several languages and is a powerful tool for natural language processing tasks.

4 Potential Benefits of AI for Higher Education Business Schools: Institution-Led, Lecturer and Student-Centred Benefits

As noted above, there is a strong and growing evidence base to demonstrate the positive impact AI and data analytics can have within education (Kuleto et al., 2021). More specifically, such impacts relate to three principal areas. These are:

- delivering convenient and personalized services
- enhancing the performance of management platforms
- offering evidence-based support for educational practice

4.1 Delivering Convenient and Personalized Services

A major benefit of AI is its ability to provide a level of personalisation and relevance in learning far beyond that which is possible within a conventional classroom model (Owoc et al., 2019). An essential objective of AI in education is to offer individualised learning guidance or assistance to students, taking into account their learning progress, preferences, and unique traits (Hwang, 2014; Hwang et al., 2020). In this way, personalised learning systems provide students with access to diverse educational materials that align with their specific learning requirements and subjects (Akgun & Greenhow, 2021). Similarly, activity recommendation systems are demonstrating increased sophistication in suggesting suitable activities for individual students based on their academic performance (Villegas-Ch et al., 2021).

Another important contribution of AI for students and their learning is related to the positive impact of these technologies when applied to large MOOC (Massive Open Online Course) datasets. Tzeng et al. (2022) for instance argue that applying AI technologies in MOOC datasets helps students to learn new skills and grasp new concepts, as well as benefit from enhanced personalised learning experience.

AI may also play a pivotal role in the assessment process. Automated marking systems are increasingly proficient in assessing student knowledge, analysing responses and offering feedback. They may automatically generate learners' assessment scores during the examination, assisting educators in gaining a better insight into students' learning progress, while simultaneously making students more aware of their learning accomplishments (Yufeia et al., 2020). As noted above, such developments are not universally welcomed and we will return to the role of AI in assessment in due course.

AI also plays a crucial role in distance education, aiming to leverage technology to bridge the gap between students and educators. In particular, it can support distance education through the integration of various intelligent systems to enhance the quality of remote learning (Kose, 2014). In a similar way, the integration of social media can be particularly effective in fostering active student engagement, collaboration, and connecting communities beyond the classroom. As noted above, one of the most obvious examples of AI developed and utilised by universities is the Chatbot (EdTech Magazine, 2020). These are now widely used to engage with individual students, for instance, to support the application process or to address routine enquiries from those already enrolled on their courses. For example, Martha, a chatbot developed by George Washington University, engages in approximately 4500 conversations each month (Kuleto et al., 2021). Chatbots, are also often present on social media platforms (Kim et al., 2019, p. 14), which can be particularly effective in fostering active student engagement, collaboration, and connections with communities beyond the classroom.

4.2 Enhancing the Performance of Management Platforms

Several studies have identified a substantial improvement in the performance of management platforms since the integration of AI-related features. For example, AI based facial authentication systems have greatly enhanced security of examinations and portal management (Khan & Alotaibi, 2020). Similarly, AI-driven routines have increasingly enhanced core administration functions, such as course scheduling and personnel data management (Li, 2021). AI technologies have also provided educational administrators and management teams with data-driven insights to strengthen their decision-making processes. Drawing from extensive data sources, AI agents can predict student attrition; identify factors influencing academic performance, and aid students in course selection (Tsai et al., 2020; Villegas-Ch et al., 2021).

4.3 Offering Evidence-Based Support for Educational Practice

The third area of significant potential for AI is in supporting the practice, decision-making and professional development of educators. As in other school activities, there is clear evidence that AI is increasingly being used to manage and complete routine tasks within a learning context. For instance, AI is now commonly used to handle day-to-day activities such as the uploading of assignments, distribution of learning materials and vocalisation of text-based questions (Gupta & Bhaskar, 2020; Huang et al., 2021). In doing so, it increases educators' space to support students with other individual and collaborative knowledge-building processes (Akgun & Greenhow, 2021; Miao et al., 2021; Limna et al., 2022).

There is also evidence though of AI adding value in higher level tasks and decisions. For instance, intelligent tutoring systems can used to recommend teaching content and learning activities that align with pedagogical requirements (Aldeman et al., 2021; Bellod et al., 2021). Indeed, in some instances, such advice may occur in real-time, through the ongoing collection and analysis of classroom data on issues such as behaviour management and questioning techniques and teachers' own responses to diagnostic assessments of their pedagogical content knowledge (Aldeman et al., 2021). AI-based models for evaluating teaching have also been constructed based on teaching data, and the 'objectivity' of such approaches has been found to both reduce teachers' defensiveness and increase their receptiveness to suggested strategies for improvement (Hu, 2021; Li & Su, 2020). Through these and other approaches, the value of AI to support professional development of educators is increasingly being recognised and valued (Gunawan et al., 2021; Lampos et al., 2021).

5 Challenges from AI for Business Schools

While AI brings many opportunities for business schools to train future leaders, there are three broad groups of concerns which must be addressed. These can be summarised as:

- Technical challenges
- Human challenges
- Curriculum challenges

5.1 Technical Challenges

Technical challenges comprise the broad set of issues which relate to the use of AI in any organisational context. One central concern relates to privacy. There are no clear policies and procedures related to who controls the data accumulated by AI, how safe that data is and how it is managed (Cheng et al., 2022). Policies and process for data governance are also critical, to control data is organized, collected, controlled, stored, used, archived, and destroyed. Establishing sound data governance relies on specific programs, well-defined policies and procedures, and clear communication from organisational leadership.

Additionally, there remains a lack of uniform standards for AI adoption in business schools, leaving its effective use dependent on several factors, including individual knowledge and analytical skills (Tong-On et al., 2021). We return to these themes subsequently in our discussion concerning the implications of AI for the curriculum content for the MBA.

A further technical challenge concerns bias. In simple terms, a major limitation of AI's use is its potential to introduce and amplify bias in a variety of forms. Correcting such bias requires human oversight and insight, which is expensive, time consuming and itself susceptible to bias in many similar ways. Within the context of business schools, bias presents particular risks in areas such as admissions processes and career advising. While some evidence of success in developing corrective measures has been reported, longer-term, sustainable solutions continue to pose significant challenges, as they require systematic approaches to addressing bias areas such as AI systems and training data (Ferrara, 2023).

5.2 Human Challenges

AI presents a plethora of challenges to students and educators, including limitations in its adaptability and flexibility, digital illiteracy, ethical considerations, lack of technological competency, resistance to learning new technology, and challenges related to time management and self-regulation. For these, and other reasons, writers such as Kabudi et al. (2021) have criticised AI as offering simplistic solutions which are inadequate for more complicated learning endeavours. In particular, they assert that AI continues to lack the adaptability and flexibility required effectively identify relevant material for courses in higher education (Owoc et al., 2021), and that academics' reliance on them to do so is premature. We will return to AI's contribution to the curation of knowledge again in the next section.

For students, there is evidence that a lack of digital literacy can also be an obstacle to effectively utilising AI (Rasheed et al., 2020). Here challenges include a lack of understanding new technology, a deficiency in technological competence, intimidation when faced with new technology, confusion and resistance in seeking appropriate online assistance, and difficulty in understanding instructions and expectations when using artificial intelligence systems (Hubbart, 2023). Such limitations have become exacerbated by the fact that the utilisation and application of new technologies such as AI is commonly premised on the expectation that students take ever-greater responsibility for their own leaning journey. Implicit in this are associated assumptions that students are capable of demonstrating greater independence, self-management, and self-organisation to navigate a variety of resources and challenges without explicit direction (while simultaneously demonstrating the necessary fortitude to avoid the potential danger of procrastination due to a lack of self-regulation skills (Chuang et al., 2018)). AI also presents significant ethical concerns for students, some of which echo challenges for educators flagged above, for instance in relation to insufficient guidelines and policies (Nguyen et al., 2023). Other concerns include data privacy, security, fairness, equity, transparency, and access to technology, all of which collectively influence students' learning experience within Business Schools.

For lecturers, McGrath et al. (2023) identify two broad categories of concern. These are broadly differentiated between those which are infrastructure based, and those which are people-centred. Many aspects of the former relate to issues discussed elsewhere in this chapter, for instance concerning responsibilities, governance and oversight. A lack of strategic planning is also a perceived weakness in many universities (McGrath et al., 2023). We will return to this and other leadership challenges subsequently in our discussion concerning the curriculum for the MBA. Many people-centred issues concern a fear amongst some academics that AI fundamentally challenges the nature of their role and threatens established ways of teaching. Some educators are also sceptical as to AI's genuine value and its ability to actually deliver the transformative benefits suggested, at least at a pan-disciplinary level. A lack of both knowledge and resources for engaging with and learning about AI is often a key factor in both regards, with pressure of time being a particular concern. A further issue for lecturers is the wider impact of technological change on the nature of students in the future. In short, advancements such as AI have considerable potential to affect the expectations and capabilities of future generations of students. At very least, subsequent generations of business students will be exposed to AI-enabled education from a very young age and therefore likely to be much more comfortable with and eager for machine-mediated education. Such students will therefore likely require a higher level of 'learning automation', which academics and learning technicians will need to be able to address, while simultaneously dealing with wider as (Stine et al., 2019).

5.3 *Curriculum Content*

The third challenge affecting business schools is how best to promote the necessary intellectual capacity to enhance and develop AI further. Aspects of this challenge include understanding complex text and concepts, formulating logical deductions, addressing complex issues, and presenting compelling arguments (Bunch, 2020). Alongside this, continually enhancing students' critical thinking and analytical skills remains an imperative. In practice, building such capacity may be best achieved by developing pedagogical approaches which utilise AI itself as one mechanism through which the concerns outlined above are considered. In such a strategy, the medium effectively becomes part of the message, and in turn helps to develop a more immersive and authentic learning experience for students and staff alike (Sollosy & McInerney, 2022). Indeed, the importance of building capacity amongst educators cannot be overstated, given that a lack of effective instructors has long been highlighted as a major contributing factor in the continued shortage of expertise in this field (Brown et al., 2011).

In the following section, we examine further how such an integrated approach may work within the context of Business School's flagship learning offer of the MBA.

6 Developing Leaders: How MBAs Can Help Prepare Leaders for the AI Age

Leadership can be defined in a variety of different ways but is commonly recognised as a *process* through which a *leader influences a follower* to pursue and achieve a specific, *aspirational vision* (Northouse, 2019; Yukl, 2013). Collectively these three elements can be viewed as the 'how', 'who' and 'why' of leadership respectively. The exact focus for and response to each of these questions is a unique construction, developed for and in light of the demands of the specific context within which leadership takes place. Thus, each leadership interaction is unique and developed to fit the specific requirements of the situation. The aim of the MBA programme therefore is to provide leaders with the intellectual capital required to navigate this process and develop the leadership approach most likely to secure the desired outcome.

While the effective introduction and utilisation of AI can in some respects be considered as 'just another' leadership challenge, the earlier discussion in this chapter demonstrates how it is characterised by several features that collectively present a variety of particularly complex challenges to leaders. Core to these are the novelty of the issue being addressed, its potential for transformational change, and the rapidity, complexity and unpredictability of the technology which is driving it (Mrowinski et al., 2020). Therefore we argue that because of these factors, AI represents the single most challenging, or 'wicked' problem (Horn & Weber, 2007), leaders will face over the next decade. However, an examination of the literature quickly reveals that while there is an appetite amongst many educators and students alike to explore AI and its implications for leaders within the MBA (e.g. Elhajjar et al., 2021; Kumar & Raman, 2022), moves to do so have to date been limited (Chen, 2022; Xu & Babaian, 2021).

In light of this, we argue that business schools need to review and redesign their MBAs programmes to ensure that the curriculum for this flagship qualification provides the intellectual capital and necessary blend of technical knowledge and business acumen to develop the 'purple' leaders (Eckerson, 2010) needed to successfully maximise the potential value of AI to their organisation. In our view, this process involves addressing the following four key questions, each of which relates to a specific dimension of the curriculum:

1. Which <u>leadership capabilities</u> will be most important for leaders in this context?
2. What <u>specialist knowledge</u> will leaders need to develop to maximise the potential of AI?
3. What are the <u>teaching and learning strategies</u> most likely to help learners develop this intellectual capital?
4. How can learners best <u>demonstrate their mastery</u> of these capacities?

Our proposed answer to each of these questions is summarised in our explanatory framework outlined in Fig. 1. This shows the curriculum as comprising four separate elements of [i] aims, [ii] content, [iii] teaching and [iv] learning strategies. While the precise nature of each is context specific, items [ii–iv] should always be

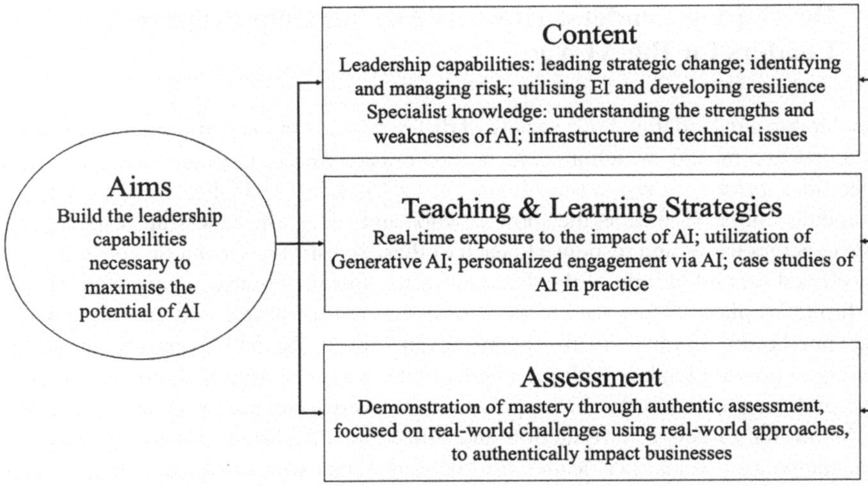

Fig. 1 Curriculum design for considering AI within MBA programmes

subservient to the overarching aim [item i] which is intended to be addressed. At the same time however, a symbiotic relationship is required between the three pathways through which this is realised to ensure that they remain both aligned and mutually informing.

The remainder of this section explores each element of this proposed curriculum in turn, drawing on relevant existing research in this field.

6.1 Leadership Capabilities Required by Leaders in Implementing AI

Research into the leadership implications of AI is patchy, with the field largely underexplored (Dixit et al., 2021; Suseno et al., 2022). Notable exceptions to this though include Peifer et al. (2022) who identify four specific domains for leaders to focus on, these being [i] managing the strategic transformation process, [ii] generating acceptance amongst stakeholders for change, [iii] establishing an appropriate culture and [iv] the human/AI interaction. Elsewhere, Tarisayi (2023) differentiates more broadly between the need for technical leadership and technical management, with the former focused on pioneering innovation and disruption, and the latter concerned with the prudent adoption and governance of technology for business benefits. In a similar way, Anghel (2023) utilises the concept of Leadership Complementary Skills to understand the specific characteristics required to oversee the introduction of AI successfully.

In broad terms, however, this and other related literature indicate that AI demands three specific forms of leadership capability, which in turn, should be central to the MBA curriculum. These are:

- Leading strategic change to address the wicked problem of AI
- Identifying and managing risk
- Utilising emotional intelligence and developing resilience.

6.1.1 AI Leading Strategic Change to Address the Wicked Problem of AI

Leaders are agents of change. Indeed, it is the focus on change which makes leadership what it is, i.e. what differentiates it from the closely related field of managerialism (Mintzberg, 1973). While a plethora of academics and business gurus alike have mused on the various dimensions of change in theory and practice, a closer inspection of models which describe the process of change (Bullock & Batten, 1985; Kotter, 1996; Lewin, 1947) shows that many essentially contain four distinct activities. These can be summarised as:

- Diagnosis—i.e. understanding the nature of the issue, potential solutions, and the resources available to deliver them
- Planning—i.e. securing commitment to pursue the desired strategy and outcomes, and detailed planning for its delivery
- Delivery—i.e. implementing the change plan, monitoring its progress and modifying this approach where necessary
- Consolidation—i.e. ensuring new ways of working are effectively integrated into the system and successfully transitioned to business as usual.

While these phases are generic and can be seen to apply in most contexts, it is also true that the introduction and utilisation of AI is a particularly challenging change for a variety of reasons. Firstly, there is no consensus across society as a whole as to what the focus for the development of AI should be and the risks and opportunities associated with this (Kietzmann & Pitt, 2020) The level of uncertainty around AI is therefore unprecedented (Mrowinski et al., 2020). Secondly, the process through which AI-related change occurs is both proactive [i.e. strategically driven] and reactive [i.e. in response to technology] (Dolata, 2009). This makes it highly unpredictable, adding to this uncertainty further. Thirdly, AI technology itself can be highly complex, opaque, and liable to behave in unpredictable ways, resulting in unforeseen outcomes (Dixit et al., 2021; Kietzmann & Pitt, 2020). The utilisation of AI also introduces a myriad of ethical implications, ranging, for example, from practical considerations such as how it should utilise personal information through to more existential questions such as its implications for people's day to day work, and the dangers of dependency on the relatively small number of organisations who develop the most advanced AI systems (Akinwalere & Ivanov, 2022; Dixit et al., 2021; Popenici & Kerr, 2017; Tarisayi, 2023). That one response to such issues is to advocate the development of yet more powerful AI systems to solve such ethical problems (allowing leaders to effectively abdicate their responsibilities for doing so), illustrates the pervasiveness of such technologies even at such a relatively early stage of their effective utilisation (De Cremer & Narayanan).

Collectively, these factors make AI a particularly difficult, or 'wicked', problem (Rittel & Webber, 1973). Here, wicked problems are differentiated from more common 'technical' problems leaders face, insofar as while they appear relatively straight-forward, closer inspective reveals them to be ill-defined, unfamiliar and containing a variety of inter-related issues which make them volatile and unpredictable (Grint, 1997). Consequently, the priority for dealing with wicked problems commonly focuses on containment or improvement, rather than resolution. Dealing with wicked problems such as leveraging the value of AI, requires a very specific approach to leadership. At its heart is working with a wide variety of stakeholders to clarify expectations, establishing agreement as to the vision and developing the processes for achieving it (Grint, 2010). Such 'adaptive leadership' (Heifetz et al., 2009) effectively requires the leader to serve the group, working as a facilitator to provide both the space and resources needed to move forward. For MBAs programmes, developing an understanding of alternative forms of leadership problems and the most appropriate strategies to address them should therefore be a focus. We return to the issue of change later in this chapter.

6.1.2 Identifying and Managing Risk

A central feature of the discourse on AI is the extent to which observers are polarised over its implications for the future. On one side, 'utopians' laud AI as promising unprecedented advances in knowledge, technology and improvements in how we live: on the other, stand 'dystopians' for whom AI presents an existential threat to humankind (Milmo, 2023). At the heart of this discussion, is the simple fact that as already noted, AI is developing at a speed and in ways which make the future difficult to predict. It is for these reasons that the leaders of the world's seven largest economies and Unesco (amongst others) have called for a system of 'guardrails' to help steer future work on AI, based on a series of democratic principles, including the protection of human rights, fundamental freedom and privacy for personal data (Unesco, 2023). The essential aim of such guardrails is to identify and manage the risk of such negative unforeseen consequences, which may result in value destruction, rather than value creation (Kietzmann & Pitt, 2020).

Within the context of the MBA then, ensuring future leaders are familiar with the principles of risk management and how these can be applied to AI is therefore critical. While a wide variety of risk models exist [e.g. FMEA, HACCP, FTA and Bowtie to name just a few], at the heart of many is the principle that managing risk involves:

- Establishing a systematic mechanism for identifying the most significant threats facing a specific activity, and
- Developing a mechanism for assessing their potential implications and in turn, identifying the most appropriate risk management strategy to adopt.

For AI, the most commonly cited risks are summarised in Table 1. We will revisit several of these themes elsewhere in this chapter.

6.1.3 Emotional Intelligence and Building Resilience

As noted above, AI is a highly disruptive technology; indeed, for some commentators, it is a cornerstone of a fourth industrial revolution, the significance of which could outstrip each of those which came before (Al Mawali et al., 2021; Lund, 2021; Mahmood & Mubarik, 2020). Such upheaval produces significant emotional challenges as it calls into question many of our taken-for-granted assumptions in core areas of our everyday lives. For example, the integration of new technologies in banking and retail in recent years has fundamentally altered how transactions are carried out in these areas (Mrowinski et al., 2020). While welcomed by some, such changes have also had unwanted consequences for others, with automation leading to isolation and disenfranchisement for some of the most vulnerable individuals in our societies.

The emotional consequences of AI will, without question, be greater for a variety of reasons. Firstly, its impact on employment and the nature of work will be more profound, both in the number of roles affected and the significance of changes likely to take place. McKinsey for example predicts that 50% of jobs will be impacted by AI (Manyika et al., 2017), with roles expected to be affected in almost all areas of the economy (Tschang & Almirall, 2021). Secondly, the nature of such changes are themselves difficult to predict but can be expected to vary markedly in scale and focus. Lytras (2023) for instance suggests AI may lead to new core business processes, enable new business models and fundamentally alter the value propositions

Table 1 Summary of the principal risks associated with AI

Risk	Description of the requirement
Ethical	Clearly define the value proposition of AI systems to ensure that they are compliant with evolving regulation and ethical guidelines, e.g. to ensure that the interests of specific groups are not prejudiced through bias
Governance	Establish comprehensive mechanisms to control and oversee the use of AI, e.g. through ensuring engagement with the most important stakeholders
Implementation	Apply recognised change management tools and processes to ensure AI systems are developed in conjunction with subject matter experts and stakeholders to meet the required performance threshold, and thereby increase the likelihood of their subsequent utilisation, e.g. by managing alternative levels of system risks
Performative	Apply rigorous methods to building AI systems, to ensure they address a clearly specified objective to a defined standard, and that limitations in performance are mitigated, e.g. the production of false positives or negatives
Security	Create strict privacy and security processes to protect AI systems from identified vulnerabilities, e.g. cyberattacks, unauthorised access and to safeguard the integrity and confidentiality of personal information.

of core business functions, while also leading to new innovation and business excellence capabilities.

For MBA directors, focusing on the emotional dimensions of leadership will be critical in developing individuals with the capabilities to lead such changes in the future. Indeed, managing the consequences of such changes will place a premium on emotional intelligence in a variety of ways. Firstly, managerial roles will not be exempt from this upheaval, as monitoring and decision-making become increasingly dependent on AI (Peifer et al., 2022). Possessing personal resilience to negotiate this transition will therefore be critical for leaders (Peifer et al., 2022). At the same time of course, displaying such characteristics in public will also demonstrate the value of emotional self-regulation to others facing similar upheaval (Goleman, 2000; Haver et al., 2014; Salovey & Mayer, 1990). Secondly, managing anxiety is a critical part of promoting successful change, both in general, and in the adoption of technology more specifically (Thatcher et al., 2007; Venkatesh, 2000). Within AI, levels of anxiety may be particularly significant. As noted above, the dystopian narrative for AI is well established and illustrated for example in ethical and legal debates as to how much trust we should place in it, for example in relation to developments such as autonomous vehicles and medical assistance devices (Hengstler et al., 2016). The fact that yet more sophisticated versions of AI is seen by some as the answer to such AI-centred ethical considerations offers little if any comfort (De Cremer & Narayanan, 2023).

Little wonder then that nearly a decade has passed since Bellini et al. (2016) identified AI anxiety as a phenomenon in its own right. For leaders, being able to address such anxiety is critical if the potential of AI is to be realised. Suseno et al. (2022) for instance found managing anxiety was essential to achieving change readiness, due to its influence on an individual's affective state [which along with cognitive and behavioural elements, is one of three domains which influence change readiness].

6.2 Specialist Knowledge

6.2.1 Understanding the Strengths and Weaknesses of AI

Possessing a sufficient understanding of the potential benefits and limitations of AI is critical if leaders are to maximise its impact. However, this is no mean feat given the wide variety of views which exist relating to it. As Stephen Hawking noted: "[AI] could be the biggest event in the history of civilization. Or the worst. We just don't know so we cannot know if we will be infinitely helped by AI or ignored by it, side-lined or conceivably destroyed by it" (Kietzmann & Pitt, 2020, p. 132). So, for leaders, understanding the intellectual, ethical and technical limitations of AI is critical if they are to successfully navigate the tightrope between blind optimism and catastrophisation in assessing its potential (Kietzmann & Pitt, 2020). For directors of MBAs, helping leaders to recognise the good from the snake oil (Courtoy &

Bastian, 2021) is therefore critical. In practice, this means providing them with the capacity to develop a sufficiently deep understanding of AI's opportunities, trends, challenges and applications to differentiate the hype from reality and critically evaluate the offer of alternative vendors (Tarisayi, 2023). It also means helping leaders to clearly recognise the limitations of their knowledge and which questions to ask of others to understand the art of the possible in their specific context (Courtoy & Bastian, 2021). One specific example of this involves providing learners with access to a wide range of real-life examples of how AI is utilised in a variety of industrial settings to assess the benefits produced, the costs incurred and the processes through which these transitions were managed. In doing so, due consideration also needs to be given to identifying instances where AI has not worked as intended, to highlight its limitations and the fact that it is not necessarily the solution for every ambition. Kietzmann and Pitt (2020) for instance note that in practice AI may include trade-offs between accuracy and interpretation, while Dixit et al. (2021) highlight a variety of potential biases which exist within AI systems, including concerning representation, measurement and aggregation.

6.2.2 Infrastructure and Technical Issues

In addition to appreciating the general strengths and weaknesses of AI, leaders also need to understand the practical implications of its utilisation. For MBAs, therefore, the goal is to develop 'purple' leaders who possess sufficient [blue] technical skills, alongside the necessary [red] leadership capabilities to take AI forward (Davenport, 2023; Eckerson, 2010). In reality, such technical knowledge may be a relatively broad-brush but should include an understanding of the technical infrastructure required to support AI, as well as the governance and oversight structures, financial resources, professional networks and specific technical expertise required to develop and oversee AI on a day-to-day basis (Tarisayi, 2023).

6.3 Teaching and Learning Strategies

Tarisayi (2023, p. 7) observes that if its full potential is to be realised, "managers must not only articulate the goals and expected outcomes of AI projects, but also lead by example in the use of AI." In the context of MBA programmes, this demands course directors give due regard to the *what* of AI as curriculum content of their programmes [i.e. the leadership capacities and additional specialist knowledge that leaders need to develop]: AI should also impact *how* learners on these programmes learn, i.e. through the delivery of content and the assessment of learning. In this way then, the message is also the medium, as studying the MBA provides future managers with a real-world, real-time exposure to the potential impact of AI. At the time of writing, relatively little consideration has been given to the impact of AI within higher education (Akinwalere & Ivanov, 2022), and indeed that which has, has often focused more on what's possible than what's already occurred. Nevertheless, AI has already clearly had a marked impact on many areas of course design and delivery within universities. Furthermore, and as is the case in all areas of society, the rapidity and significance of such changes is only likely to increase (Akinwalere & Ivanov, 2022; Popenici & Kerr, 2017).

AI has long been successfully established within many universities' libraries' search facilities. Therefore, by stealth, it has come to perform a fundamental role as both gatekeeper and curator of knowledge (Akinwalere & Ivanov, 2022). At the same time, data mining techniques and natural language processing chatboxes such as Chat GPT are playing increasingly important roles in identifying the focus for learning. As noted earlier, AI also has significant potential to directly impact the learning experiences of individual students on the programme, by influencing their engagement with content covered in a course (Zuboff, 2019). For example, AI can personalise the path a learner takes through a module, tailoring content and learning activities to identified individual needs and preferences. For educators, such AI-based techniques provide a highly efficient mechanism for identifying students' engagement with alternative material. Here AI may be especially powerful in identifying at-risk students, and, developing strategies for providing the additional support required (Tarisayi, 2023). At the same time, of course, universities need to be cognizant of the support lecturers may need if they are to develop the capacity to take advantage of such opportunities, for instance in terms of the technical skills required and in dealing with the potential threats it poses to their identity as educators (Bayne, 2015; Popenici & Kerr, 2017).

6.4 *Demonstration of Mastery*

The final area where MBA directors need to consider AI is in how learners demonstrate their mastery of learning.

AI is now well established in many higher education institutions as a mechanism for managing assessment of learning—for example automation has long been used to rapidly and efficiently score simple tests. However, in recent years AI has also become increasingly powerful in assessing the quality of written material, including the longer-form essays and reports often adopted on MBAs (Zuboff, 2019). At one level, plagiarism-detection systems have been embedded in many university assessment systems for several years. However, AI is increasingly able to provide additional support through the provision of detailed and personalised feedback for students on the work submitted (Tarisayi, 2023). Nevertheless, the use of AI in this area remains both under-explored and contentious. Kumar and Raman (2022) for instance note that while 71% of students believe AI should be used in teaching and learning processes, only 33% felt it should be used in examination processes. Similarly, discussions on integrating the use of AI into MBAs and other masters-level content have largely focused on AI as a focus for assessment, rather than a mechanism for completing assessment itself (e.g. Lytras, 2023).

In light of this, we believe it is helpful to revisit the wider discourse on the purpose of assessment itself and in particular, its contribution to preparing students the utilise AI professionally and in their life beyond the classroom. In doing we note that the last forty years has seen a growing appreciation of the role assessment can play as an active part of the learning process (Sambell et al., 2013). Core to this is a recognition that for many students, 'traditional' forms of assessment often represent hurdles to jump over and are an adjunct to the process of learning rather than a mechanism for embedding it further (Ramsden, 2003). To expediently address this requirement, students may frequently engage in shallow learning to build the capacity to effectively 'mimic' the observed requirements of the academic system (Ramsden, 2003). However, such superficial learning does little to build the capabilities required in their post-university careers (Biemans et al., 2004). Within the context of the issues discussed in this chapter, such abstract and shallow knowledge will leave MBA graduates ill-equipped to maximise the potential of AI within their organisations.

In response to such concerns, interest in more 'authentic' forms of assessment (Torrance, 1994) has increased markedly, not least because such strategies are both more meaningful and engaging for students, and in turn encourage 'deep learning', which can more readily be utilised in a professional context. Within the context of MBA programmes, there is considerable potential to develop a wide variety of authentic assessment strategies which require learners to demonstrate their mastery of the implications of AI for leaders' issues in authentic ways, and which in turn, motivate, inspire and build their capacity to address such issues subsequently throughout their career (Davison, 2011). Firstly, authentic assessment commonly explores real-world contexts, utilising real-world techniques, to address real-world challenges (Sambell et al., 2013). Within the context of AI, this may mean addressing any of the substantive themes outlined above, for instance, leading strategic change, managing risk, addressing ethical considerations or establishing the conditions for maximising the potential impact of AI on the learner's organisation. Working in partnership with businesses themselves is critical in this, both in terms of ensuring the authenticity of such work but also maximising its potential impact (Wisniewski, 2020; Elhajjar et al., 2021).

Authentic assessment also commonly employs a broader range of media to help learners develop a wider range of skills than normally associated with traditional strategies of essays and exams (Spendlove, 2009). For the MBA, utilising techniques which are valued within industry not only enables learners to demonstrate their mastery of the subject; they also provide a safe space within which wider capabilities can be developed and tested further. For example, an assessment that requires students to produce a short presentation, targeted at senior managers, in which they analyse a specific issue related to AI and offer recommendations for a specific business can develop horizon scanning, strategic thinking, and communication skills. If the student is expected to deliver this presentation within a wider group context, quick thinking, adaptability and resilience to cope under pressure are also developed further. For MBA learners already working in senior positions, there is therefore potential to connect assessment with their substantive careers in a very tangible and impactful way; for example, requiring a learner to produce a strategy and related action plan for addressing a specific AI dimension relevant to their work has the potential to promote learning so profound it becomes almost visceral. In this way then, assessment not only demonstrates the achievement of identified learning outcomes; it actively serves to develop digital leaders, able to promote innovation and collaboration, and in turn leverage the potential of AI to enhance the competitiveness of their business in the future (Petrucci & Rivera, 2018).

7 Conclusions

The rapid development of AI is impacting the whole society, from our individual lives to our professional lives, to all industries, sectors and organisations worldwide. While the pace of change is increasing, different industries and organisations implement AI in different ways and speeds, resulting in complexity, diversity and uncertainty.

The expansion of AI brings unprecedented opportunity. However, it also presents a variety of challenges, for example in relation to ethical concerns, regulations, policies and frameworks, and the fear of job displacements. Of course, Business Schools are not exempt from this. Indeed, they face the unique challenge of both dealing with AI themselves while also enabling others to both understand and utilise this disruptive power to best effect.

In this chapter, we have discussed how Business Schools can develop their MBA to address these challenges. More specifically we have outlined a model for an integrated curriculum design, which utilised its curriculum content, teaching and learning strategies, and assessment techniques to develop the 'Purple Leaders' needed to thrive in such an uncertain future. In this context, such leaders possess a unique form of leadership capital, which blends the adequate understanding of the AI itself [in terms of its strengths and limitations, and associated technical challenges], and the generic leadership capabilities to lead the organisational transformation required in order to utilise it. For business schools, this means demonstrating in practice how AI can be successfully utilised in practice, to improve the delivery of MBAs and students' experiences of them, while also providing sufficient emotional and technical support to lecturers to enable them to develop the resilience necessary to cope with the implications of such transformational change in their own contexts.

Future prosperity demands continuous adaptation to an uncertain and ever-evolving landscape. For business schools, this means investing further in research into AI and its implications, while demonstrating openness to implementing the lessons this reveals. In practice, this requires business schools adopt a collaborative approach and demonstrate the ability to leverage expertise to implement, adapt and convert AI-driven solutions into growth and innovation opportunities for business schools.

References

Accenture. (2022). *What is artificial intelligence (AI) & why is it important?* Accenture. [online].

Adeshola, I., & Adepoju, A. P. (2023). The opportunities and challenges of ChatGPT in education. *Interactive Learning Environments*, 1–14.

Akgun, S., & Greenhow, C. (2021). Artificial intelligence in education: Addressing ethical challenges in K-12 settings. *AI and Ethics*, 1–10.

Akinwalere, S. N., & Ivanov, V. (2022). Artificial intelligence in higher education: Challenges and opportunities. *Border Crossing, 12*(1), 1–15. https://doi.org/10.33182/bc.v12i1.2015

Al Mawali, N. R., Al Lawati, A. M., & Ananda, S. (2021). *Fourth industrial revolution and business dynamics: Issues and implications* (1st ed.). Palgrave Macmillan. https://doi.org/10.1007/978-981-16-3250-1

Aldeman, N. L. S., de Sá Urtiga Aita, K. M., Machado, V. P., da Mata Sousa, L. C. D., Coelho, A. G. B., da Silva, A. S., da Silva Mendes, A. P., de Oliveira Neres, F. J., & do Monte, S. J. (2021). A platform for teaching glomerulopathies using machine learning. *BMC Medical Education, 21*(1), 248.

Anghel, D. (2023). New perspectives for human and artificial intelligence interactions for leadership e-recruitment. *Societies, 13*(3), 55.

Barari, A., & Kumar, A. (2023). What is ChatGPT? Best uses and limitations of the brand-new tool. *Emeritus Online Courses*. Retrieved December 29, 2023, from https://emeritus.org/blog/ai-ml-what-is-chatgpt/

Bayne, S. (2015). Teacherbot: Interventions in automated teaching. *Teaching in Higher Education, 20*(4), 455–467. https://doi.org/10.1080/13562517.2015.1020783

Bellini, C. G. P., Isoni Filho, M. M., de Moura Junior, P. J., & de Cássia de Faria Pereira, R. (2016). Self-efficacy and anxiety of digital natives in face of compulsory computer-mediated tasks: A study about digital capabilities and limitations. *Computers in Human Behavior, 59*, 49–57. https://doi.org/10.1016/j.chb.2016.01.015

Bellod, H. C., Ramón, V. B., Fernández, E. C., & Luján, J. F. G. (2021). Analysis of stress and academic-sports commitment through self-organizing artificial neural networks. *Retos: nuevas tendencias en educación física, deporte y recreación, 42*, 136–144.

Biemans, H., Nieuwenhuis, L., Poell, R., Mulder, M., & Wesselink, R. (2004). Competence-based VET in the Netherlands: Background and pitfalls. *Journal of Vocational Education & Training, 56*(4), 523–538. https://doi.org/10.1080/13636820400200268

Brown, B., Chui, M., & Manyika, J. (2011). Are you ready for the era of 'big data'. *McKinsey Quarterly, 4*(1), 24–35.

Bullock, R., & Batten, D. (1985). It's just a phase we're going through. *Group and Organizational Studies, 10*, 383–412.

Bunch, K. J. (2020). State of undergraduate business education: A perfect storm or climate change? *The Academy of Management Learning and Education, 19*(1), 81–98.

Chen, L. (2022). Current and future artificial intelligence (AI) curriculum in business school: A text mining analysis. *Journal of Information Systems Education, 33*(4), 416–426.

Chen, L., Chen, P., & Lin, Z. (2020). Artificial intelligence in education: A review. *IEEE Access, 8*, 75264–75278.

Cheng, X., Su, L., Luo, X., Benitez, J., & Cai, S. (2022). The good, the bad, and the ugly: Impact of analytics and artificial intelligence-enabled personal information collection on privacy and participation in ridesharing. *European Journal of Information Systems, 31*(3), 339–363.

Chuang, H. H., Weng, C. Y., & Chen, C. H. (2018). Which students benefit most from a flipped classroom approach to language learning? *British Journal of Educational Technology, 49*(1), 56–68.

Courtoy, J. C., & Bastian, N. D. (2021). Three things leaders need to know before investing in artificial intelligence. *Phalanx (Alexandria), 54*(1), 32–37.

Davenport, T. (2023). Purple people at the heart of cognitive tech. *Wall Street Journal.*

Davison, G. (2011). *Investigating the relationships between authentic assessment and the development of learner autonomy*. Northumbria University.

De Cremer, D., & Narayanan, D. (2023). On educating ethics in the AI era: Why business schools need to move beyond digital upskilling, towards ethical upskilling. *AI and Ethics*, 1–5.

Dixit, A., Quaglietta, J., & Gaulton, C. (2021). Preparing for the future: How organizations can prepare boards, leaders, and risk managers for artificial intelligence. *Healthcare Management Forum, 34*(6), 346–352. https://doi.org/10.1177/08404704211037995

Dolata, U. (2009). Technological innovations and sectoral change. Transformative capacity, adaptability, patterns of change: An analytical framework. *Research Policy, 38*(6), 1066–1076. https://doi.org/10.1016/j.respol.2009.03.006

Duan, N., Liu, L. Z., Yu, X. J., Li, Q., & Yeh, S. C. (2019). Classification of multichannel surface-electromyography signals based on convolutional neural networks. *Journal of Industrial Integration and Management, 15*, 201–206.

Eckerson, W. (2010). "Purple People": The key to BI success. *TDWI Blog*. Retrieved December 22, 2023, from https://tdwi.org/Blogs/TDWI-Blog/2010/04/Purple-People.aspx

EdTech Magazine. (2020). Retrieved October 2023, from https://edtechmagazine.com/higher/article/2020/02/5-year-vision-artificial-intelligence-higher-ed

Educase Horizon Report. (2021). *Teaching and Learning Edition*. [Online]. Retrieved October 2023, from https://library.educause.edu/resources/2021/4/2021-educause-horizon-report-teaching-and-learning-edition

Elhajjar, S., Karam, S., & Borna, S. (2021). Artificial intelligence in marketing education programs. *Marketing Education Review, 31*(1), 2–13.

Ferrara, E. (2023). *The butterfly effect in artificial intelligence systems: Implications for ai bias and fairness*. Available at SSRN 4614234.

Foffano, F., Scantamburlo, T., & Cortés, A. (2023). Investing in AI for social good: An analysis of European national strategies. *AI & Society, 38*(2), 479–500.

Goleman, D. (2000). Leadership that gets results. *Harvard Business Review*, 78–90.

Grint, K. (1997). *Leadership: Classical, contemporary, and critical approaches. Oxford management readers*.

Grint, K. (2010). *Leadership: A very short introduction* (Vol. 237). Oxford University Press.

Gunawan, K. D. H., Liliasari, L., Kaniawati, I., & Setiawan, W. (2021). Implementation of competency enhancement program for science teachers assisted by artificial intelligence in designing HOTS-based integrated science learning. *Jurnal Penelitian dan Pembelajaran IPA, 7*(1), 55–65.

Gupta, K. P., & Bhaskar, P. (2020). Inhibiting and motivating factors influencing teachers' adoption of AI-based teaching and learning solutions: Prioritization using analytic hierarchy process. *Journal of Information Technology Education. Research, 19*, 693.

Haderer, B., & Ciolacu, M. (2022). Education 4.0: Artificial intelligence assisted task-and time planning system. *Procedia Computer Science, 200*, 1328–1337.

Haver, A., Akerjordet, K., & Furunes, T. (2014). Wise emotion regulation and the power of resilience in experienced hospitality leaders. *Scandinavian Journal of Hospitality and Tourism, 14*(2).

Heifetz, R., Grashow, A., & Linsky, M. (2009). *The practice of adaptive leadership*. Harvard Business School Publishing.

Hengstler, M., Enkel, E., & Duelli, S. (2016). Applied artificial intelligence and trust—The case of autonomous vehicles and medical assistance devices. *Technological Forecasting & Social Change, 105*, 105–120. https://doi.org/10.1016/j.techfore.2015.12.014

Hirsh-Pasek, K., & Blinkoff, E. (2023). ChatGPT: Educational friend or foe? *Brookings*. Retrieved December 29, 2023, from https://www.brookings.edu/blog/education-plus-development/2023/01/09/chatgpt-educational-friend-or-foe/

Horn, R., & Weber, R. (2007). *New tools for resolving wicked problems* (p. 31). MacroVu Inc and Strategy Kinetics.

Hu, J. (2021). Teaching evaluation system by use of machine learning and artificial intelligence methods. *International Journal of Emerging Technologies in Learning (iJET), 16*(5), 87–101.

Huang, J., Shen, G., & Ren, X. (2021). Connotation analysis and paradigm shift of teaching design under artificial intelligence technology. *International Journal of Emerging Technologies in Learning (iJET), 16*(5), 73–86.

Hubbart, J. A. (2023). Organizational change: The challenge of change aversion. *Administrative Sciences, 13*(7), 162.

Hwang, G. J. (2014). Definition, framework and research issues of smart learning environments - A context-aware ubiquitous learning perspective. *Smart Learning Environments, 1*(1), 1–14, No. 4.

Hwang, G. J., Xie, H., Wah, B. W., & Gašević, D. (2020). Vision, challenges, roles and research issues of artificial intelligence in education. *Computers and Education: Artificial Intelligence, 1*, 100001.

Kabudi, T., Pappas, I., & Olsen, D. (2021). AI-enabled adaptive learning systems: A systematic mapping of the literature. *Computers and Education: Artificial Intelligence, 2*, 100017.

Khan, F., & Alotaibi, R. (2020). Design and implementation of a computerized user authentication system for e-Learning. *International Journal of Emerging Technologies in learning (iJET), 15*(9), 4–18.

Kietzmann, J., & Pitt, L. F. (2020). Artificial intelligence and machine learning: What managers need to know. *Business Horizons, 63*(2), 131–133. https://doi.org/10.1016/j.bushor.2019.11.005

Kim, N. Y., Cha, Y., & Kim, H. S. (2019). Future English learning: Chatbots and artificial intelligence. *Multimedia-Assisted Language Learning, 22*(3), 32–53.

Kose, U. (2014). *Artificial intelligence applications in distance education*. IGI Global.

Kotter, J. (1996). *Leading change*. Harvard Business School Press.

Kuleto, V., Ilić, M., Dumangiu, M., Ranković, M., Martins, O. M., Păun, D., & Mihoreanu, L. (2021). Exploring opportunities and challenges of artificial intelligence and machine learning in higher education institutions. *Sustainability, 13*(18), 10424.

Kumar, P., Chauhan, S., & Awasthi, L. K. (2023). Artificial intelligence in healthcare: Review, ethics, trust challenges & future research directions. *Engineering Applications of Artificial Intelligence, 120*, 105894.

Kumar, V. R., & Raman, R. (2022, March). Student perceptions on artificial intelligence (AI) in higher education. In *2022 IEEE Integrated STEM Education Conference (ISEC)* (pp. 450–454). IEEE.

Lampos, V., Mintz, J., & Qu, X. (2021). An artificial intelligence approach for selecting effective teacher communication strategies in autism education. *Science of Learning, 6*(1), 25.

Lewin, K. (1947). Group decisions and social change. In T. Newcomb & E. Harley (Eds.), *Readings in social psychology*. Henry Holt.

Li, M., & Su, Y. (2020). Evaluation of online teaching quality of basic education based on artificial intelligence. *International Journal of Emerging Technologies in Learning (iJET), 15*(16), 147–161.

Li, Q. (2021). The use of artificial intelligence combined with cloud computing in the design of education information management platform. *International Journal of Emerging Technologies in Learning (iJET), 16*(5), 32–44.

Limna, P., Jakwatanatham, S., Siripipattanakul, S., Kaewpuang, P., & Sriboonruang, P. (2022). A review of artificial intelligence (AI) in education during the digital era. *Advance Knowledge for Executives, 1*(1), 1–9.

Lund, B. (2021). The Fourth Industrial Revolution. *Information Technology and Libraries (Online), 40*(1), 1–4. https://doi.org/10.6017/ital.v40i1.13193

Lytras, M. (2023, June). Active and transformative learning for the integration of artificial intelligence courses for business school students. In *Conference Proceedings. The Future of Education 2023*.

Mahmood, T., & Mubarik, M. S. (2020). Balancing innovation and exploitation in the fourth industrial revolution: Role of intellectual capital and technology absorptive capacity. *Technological Forecasting & Social Change, 160*, 120248. https://doi.org/10.1016/j.techfore.2020.120248

Manyika, J., Lund, S., Chui, M., Bughin, J., Woetzei, J., Batra, P., Ryan, K., & Sanghvi, S. (2017). *Jobs lost, jobs gained: Workforce transition in a time of automation*. https://www.mckinsey.com/~/media/BAB489A30B724BECB5DEDC41E9BB9FAC.ashx

McGrath, C., Pargman, T. C., Juth, N., & Palmgren, P. J. (2023). University teachers' perceptions of responsibility and artificial intelligence in higher education-An experimental philosophical study. *Computers and Education: Artificial Intelligence, 4*, 100139.

Miao, F., Holmes, W., Huang, R., & Zhang, H. (2021). *AI and education: A guidance for policymakers*. UNESCO Publishing.

Milmo, D. (2023, October 24). Hope or horror? The great AI debate dividing its pioneers. *The Guardian*. https://www.theguardian.com/technology/2023/oct/24/hope-or-horror-the-great-ai-debate-dividing-its-pioneers

Mintzberg, H. (1973). *The nature of managerial work*. Harper & Row.

Mrowinski, B., Brougham, D., & Tappin, D. C. (2020). Managers' perceptions of artificial intelligence and automation: Insights into the future of work. *New Zealand Journal of Employment Relations, 44*(3), 76–91.

Nguyen, A., Ngo, H. N., Hong, Y., Dang, B., & Nguyen, B. P. T. (2023). Ethical principles for artificial intelligence in education. *Education and Information Technologies, 28*(4), 4221–4241.

Northouse, P. G. (2019). *Leadership: Theory and practice* (2nd ed.). Sage Publications.

Nunavath, V., & Goodwin, M. (2018). The role of artificial intelligence in social media big data analytics for disaster management-initial results of a systematic literature review. In *5th International Conference on Information and Communication Technologies for Disaster Management (ICT-DM)* (pp. 1–4). IEEE.

Owoc, M., Sawicka, A., & Weichbroth, P. (2021). Artificial intelligence technologies in education: Benefits, challenges and strategies of implementation. In Ue.wroc.pl. Available at: http://www.ue.wroc.pl/en/

Owoc, M. L., Sawicka, A., & Weichbroth, P. (2019). Artificial intelligence technologies in education: Benefits, challenges and strategies of implementation. In *IFIP International Workshop on Artificial Intelligence for Knowledge Management* (pp. 37–58).

Peifer, Y., Jeske, T., & Hille, S. (2022). Artificial intelligence and its impact on leaders and leadership. *Procedia Computer Science, 200*, 1024–1030. https://doi.org/10.1016/j.procs.2022.01.301

Petrucci, T., & Rivera, M. (2018). Leading growth through the digital leader. *Journal of Leadership Studies, 12*(3), 53–56.

Popenici, S. A. D., & Kerr, S. (2017). Exploring the impact of artificial intelligence on teaching and learning in higher education. *Research and Practice in Technology Enhanced Learning, 12*(1), 13–22. https://doi.org/10.1186/s41039-017-0062-8

Ramsden, P. (2003). *Learning to teach in higher education* (2nd ed.). RoutledgeFalmer.

Rasheed, R., Kamsin, A., & Abdullah, N. (2020). Challenges in the online component of blended learning: A systematic review. *Computers & Education, 144*, 103701. https://doi.org/10.1016/j.compedu.2019.103701

Rittel, H., & Webber, M. (1973). Dilemmas in a general theory of planning. *Policy Sciences, 4*, 155–169.

Salovey, P., & Mayer, J. (1990). Emotional intelligence. *Imagination, Cognition, and Personality, 9*(3), 185–211.

Sambell, K., McDowell, L., & Montgomery, C. (2013). In L. McDowell & C. Montgomery 1962 (Eds.), *Assessment for learning in higher education*.

Sollosy, M., & McInerney, M. (2022). Artificial intelligence and business education, what should be taught. *International Journal of Management Education (Elsevier Science), 20*(3).

Spendlove, D. (2009). Putting assessment for learning into practice.

Stine, J., Trumbore, A., Woll, T., & Sambucetti, H. (2019). Implications of artificial intelligence on business schools and lifelong learning. *Academic leadership group*. Retrieved December 28, 2023, from https://www.embac.org/pdf/research/2020.09/ai-final-report-11.04.19-2.pdf

Suseno, Y., Chang, C., Hudik, M., & Fang, E. S. (2022). Beliefs, anxiety and change readiness for artificial intelligence adoption among human resource managers: The moderating role of high-

performance work systems. *International Journal of Human Resource Management, 33*(6), 1209–1236. https://doi.org/10.1080/09585192.2021.1931408
Tarisayi, K. S. (2023). Strategic leadership for responsible artificial intelligence adoption in higher education. *CTE Workshop Proceedings*. https://doi.org/10.55056/cte.616
Thatcher, J. B., Loughry, M. L., Lim, J., & McKnight, D. H. (2007). Internet anxiety: An empirical study of the effects of personality, beliefs, and social support. *Information & Management, 44*(4), 353–363. https://doi.org/10.1016/j.im.2006.11.007
Tong-On, P., Siripipatthanakul, S., & Phayaphrom, B. (2021). The implementation of business intelligence using data analytics and its effects towards on performance in the hotel industry in Thailand. *International Journal of Behavioral Analytics, 1*(2), 1–16.
Torraco, R. J. (2016). Writing integrative literature reviews: Using the past and present to explore the future. *Human Resource Development Review, 15*(4), 404–428.
Torrance, H. (1994). *Evaluating authentic assessment: Problems and possibilities in new approaches to assessment*. Open University Press.
Tsai, S. C., Chen, C. H., Shiao, Y. T., Ciou, J. S., & Wu, T. N. (2020). Precision education with statistical learning and deep learning: A case study in Taiwan. *International Journal of Educational Technology in Higher Education, 17*, 1–13.
Tschang, F. T., & Almirall, E. (2021). Artificial intelligence as augmenting automation: Implications for employment. *Academy of Management Perspectives, 35*(4), 642–659.
Tzeng, J. W., Lee, C. A., Huang, N. F., Huang, H. H., & Lai, C. F. (2022). MOOC evaluation system based on deep learning. *International Review of Research in Open and Distributed Learning, 23*(1), 21–40.
Unesco. (2023, May 22). UNESCO supports G7 leaders calling for 'AI guardrails.' *Unesco*. https://www.unesco.org/en/articles/unesco-supports-g7-leaders-calling-ai-guardrails?TSPD_101_R0=080713870fab20000d91aefea2f9d5919adfc044c9338a4211dafe0099cf68ee3f110ed16f449ba30807b4f3661430000496e9175bb77341e324434962c631285a1074dcd50c8d72d4df43a-31ca445a673aa569f539d111be9506172f2da3a98
Venkatesh, V. (2000). Determinants of perceived ease of use: Integrating control, intrinsic motivation, and emotion into the technology acceptance model. *Information Systems Research, 11*(4), 342–365. https://doi.org/10.1287/isre.11.4.342.11872
Villegas-Ch, W., García-Ortiz, J., & Sánchez-Viteri, S. (2021). Identification of the factors that influence university learning with low-code/no-code artificial intelligence techniques. *Electronics, 10*(10), 1192.
Vollero, A., Sardanelli, D., & Siano, A. (2023). Exploring the role of the Amazon effect on customer expectations: An analysis of user-generated content in consumer electronics retailing. *Journal of Consumer Behaviour, 22*(5), 1062–1073.
Wisniewski, H. S. (2020). What is the business with AI? Preparing future decision makers and leaders. *Technology & Innovation, 21*(4), 1–14.
Xu, J. J., & Babaian, T. (2021). Artificial intelligence in business curriculum: The pedagogy and learning outcomes. *The International Journal of Management Education, 19*(3), 100550.
Yufeia, L., Salehb, S., Jiahuic, H., & Syed, S. M. (2020). Review of the application of artificial intelligence in education. *Integration (Amsterdam), 12*(8), 1–15.
Yukl, G. (2013). *Leadership in organizations* (8th ed.). Pearson Education.
Zhang, C., & Lu, Y. (2021). Study on artificial intelligence: The state of the art and future prospects. *Journal of Industrial Information Integration, 23*, 100–224.
Zuboff, S. (2019). *The age of surveillance capitalism: The fight for the future at the new frontier of power*. Profile Books.

The Application of Artificial Intelligence in Recruitment, Training and Employee Onboarding HR Practices

Pedro Cuesta-Valiño, Sergey Kazakov, Blanca García Henche, and Estela Núñez-Barriopedro

Abstract This chapter zooms in on the facets of proliferating Artificial Intelligence (AI) deployment in recruitment, training, and employee onboarding, which are significant procedures for the domain of Human Resources Management (HRM). At the current stage of its evolution, AI implies a synthesis of machine learning, data science, computer hardware and communication technologies. Grounded in this synthesis, AI revolutionises HR processes that have commonly exploited a tangible amount of organisational resources until recently. The chapter considers the specifics of AI utilisation in scoped HRM processes, presents real-world case studies and reveals practical insights applicable to organisations considering the adoption of AI-powered solutions to successfully upgrade the execution of their recruitment, training, and employee onboarding procedures to a new level of efficiency.

1 Introduction

Artificial Intelligence (AI) is referred to as technology aimed at the automation of human activities by developing and training software applications and hardware devices (Kolbjørnsrud et al., 2016). Researchers indicate that at present AI is fuelling momentum in the shift towards the rapid evolution of Human Resource Management (HRM) (Weitzman, 2022). Combined with advanced Information and Communication Technologies (ICT), AI transforms HRM processes and makes decision-making in HRM more efficient, thus, improving organisational performance (Mikalef & Gupta, 2021). AI usage proliferates in standard HRM routines and embraces all its facets to achieve excellent results. The core benefit of AI deployment relates to its capability to provide complete automation and integration

P. Cuesta-Valiño · S. Kazakov (✉) · B. G. Henche · E. Núñez-Barriopedro
University of Alcalá, Faculty of Economics, Business and Tourism, Department of Economics and Business, Madrid, Spain
e-mail: pedro.cuesta@uah.es; sergey.kazakov@uah.es; blanca.garcia@uah.es; estela.nunezb@uah.es

M. T. Del Val Núñez et al. (eds.), *Artificial Intelligence and Business Transformation*, Contributions to Management Science,
https://doi.org/10.1007/978-3-031-58704-7_13

of all HRM processes and procedures, putting them in one pipeline (Pan & Froese, 2023). Organisations that adopt AI solutions for HRM receive an immediate return on investment by increasing HR department performance, increasing the job satisfaction of HR managers, faster induction of new employees, and creating a positive employer brand image in the labour market thanks to the seamless interaction with candidates (Black & Esch, 2020).

Employee recruitment, training and onboarding are paramount procedures of HRM. These procedures help seek and recruit the best candidates for open positions, ensure efficient integration and functioning, maintain business and organisational performance, and achieve the organisation's global business goals (Becker & Bish, 2021). Finding, selecting and inducting qualified and appropriate candidates to fill the open positions in an organisation is a critical task of the HRM department. This detailed workflow includes all the necessary steps, from developing job descriptions, comprising the required skills and expertise for the vacancy, to onboarding and training a new employee (Davila & Pina-Ramirez, 2018). On top of that, the recruitment procedure also involves attracting and hiring qualified professionals who require time and effort from the HR manager to persuade the best candidates to accept a job offer from the organisation (Van Esch et al., 2019). The recruitment process aims to increase recruitment efficiency, cost-effectiveness, scalability, and alignment with corporate objectives and statutory requirements (Johnson et al., 2020).

Until recently, HR managers and recruiters have spent a great deal of effort searching for relevant information and screening potential candidates by manually browsing various specialised online platforms. Moreover, they used to spend much time scheduling interviews and induction-orientation training for new employees. With limited resources available, HR departments should reach a vast community of potential candidates to find and hire suitable employees for their organisation (Bingham et al., 2002). It can take weeks, months, or even years to study CVs, host endless interviews, and negotiate remuneration and fringe benefits to hire an appropriate candidate for a position which needs to be filled as soon as possible. Besides, when organisations needed to engage in mass hiring, they were supposed to hire additional HR managers to handle these complex projects. Furthermore, decision-making in hiring tended to be entirely subjective to the specific HR manager, thus engendering a low degree of adequate candidate evaluation (Chowdhury et al., 2023).

By utilising literature review and case study research methodologies, this chapter examines the utilisation of AI solutions for several basic HRM tasks, including recruitment, training, and employee onboarding. It documents the state-of-the-art AI-empowered HRM solutions helping organisations in seeking, screening, selecting, attracting and hiring the top talents available in the labour market to securing a rapid, smooth and straightforward training and onboarding of any newly hired employee in their new role. By employing AI-based solutions, organisations can hire the best professionals for the job, providing better results in employee hiring and onboarding.

This chapter begins with an in-depth analysis of AI-powered recruitment and selection strategies, unravelling the sophisticated ways intelligent systems enhance

HRM efficiency and objectivity in talent acquisition. After this, the chapter depicts several case studies of successful AI system deployment in HRM processes, exhibiting their robust implementation based on real-life examples. Furthermore, the present chapter explores applications of AI-driven onboarding and training solutions, unveiling the efficiency of AI applications for smooth employee integration into a new job and ongoing Training and Development (T&D) in the organisation. Finally, this chapter concludes with envisaging AI deployment in HRM processes in the future, evincing the call for organisations to steer and speed up the process of AI adoption in HRM processes. Such an effort can ensure an organisation's future success in the volatile HRM realm and the contemporary business environment.

2 AI-Powered Recruitment and Selection

In times of global change, the modern organisation of HRM requires a recruitment process to utilise more robust and faster techniques for applicant selection, recruiting and onboarding. AI in recruitment process deployment has changed and streamlined the task of hiring personnel in many organisations all over the world. AI recruitment applications are built with the help of advanced data science methods comprising machine and deep learning, Natural Language Processing (NLP) frameworks, and sophisticated neural network modelling methods (Chowdhury et al., 2023). These algorithms simplify and speed up the recruitment and hiring process, making them more productive, objective, and focused on the hiring task, organisational needs, and candidate suitability (Johnson et al., 2020). To achieve better HRM performance, organisations utilise AI to a significant degree in the recruitment process by sourcing and, screening candidates, interview scheduling and new employee onboarding. However, the level of the available job position in an organisation and the non-linear nature of these processes must be considered by HR managers. In this context, relying on AI-based recruitment solutions works best for mass hiring or low- and middle-level positions in organisations. Simultaneously, recruitment for top and executive vacancies requires more human input from HR managers.

2.1 AI-Powered Applicant Tracking Systems for Efficient Candidate Sourcing, Screening and Scheduling

According to 52% of surveyed HR managers, the most challenging facet of recruitment is finding suitable candidates from a vast range of available applicants (ATS Consulting Services, 2023). AI recruitment solutions employed by organisations can help to rationalise job candidates' seeking process. At the initial stage of employee recruitment, the AI-driven candidate sourcing and screening procedure commonly comprises AI applications of various types suitable for a concrete task in the process (Weitzman, 2022). In this vein, organisations start the recruitment process by administering Intelligent Candidate Tracking Systems (ICTS), which can work with external CV databases and job applications received after distributing job postings. To function appropriately, AI-empowered ICTS needs to receive all the necessary information, including job description, required experience and skills, keywords or tags to help to better match candidates. ICTS apply NLP-based models to find a pool of appropriate candidates automatically (Fig. 1). Without input from the HR manager, it may rank the potential candidates based on their suitability for the organisation and proceeds to the next stage of the recruitment process, known as screening.

In the process of screening potential selected candidates, HR managers may use a set of various ICTS solutions. They comprise chatbots or automated text and voice messaging tools which communicate with applicants chosen for an initial screening of their experience, provide information about the open position, answer possible questions from the applicants, filter out a candidate or set an appointment with the recruiter (Kaplan & Haenlein, 2019; Jobvite, 2021). Several ICTS solutions can host and analyse the results of one-way recorded digitised interviews as a part of the candidates' screening process (Koechling et al., 2023). A candidate connects for the interview as they would for a regular conference call using his computer and camera and responds to the questions asked by the machine. ICTS solutions analyse the results and recommend the HR manager to meet with the selected candidates in the following interview round in person (*The Muse*, 2016).

Supplementary ICTS solutions include online qualification tests, while other AI applications implement gamification for candidate screening. HR managers routinely use several types of AI games in the screening process. For instance, applicants for programmer positions may complete an online quiz, coding test, or solve

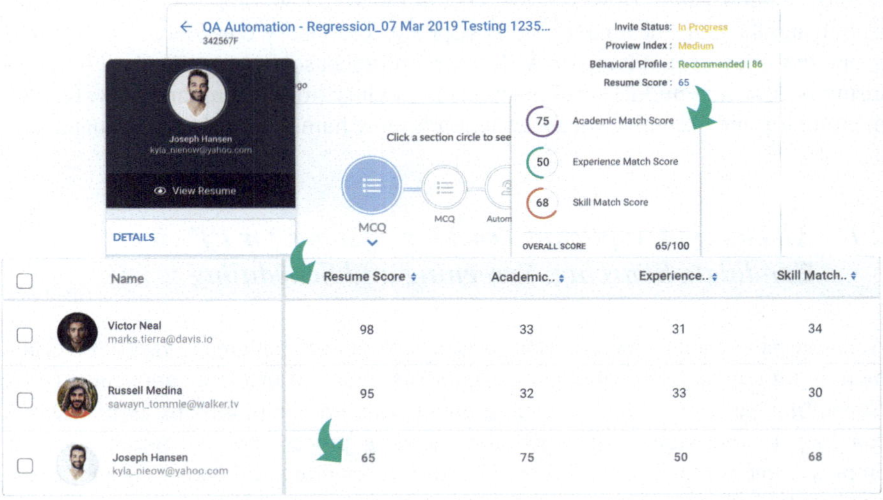

Fig. 1 Example of ICTS selecting a pool of potential candidates. Source: Gomes (2023)

online coding exercises. Puzzle games are utilised for candidates applying for various open jobs to determine the candidate's intellect and ability to resolve challenges arising during the job. Simulation games have become popular due to the possibility of checking candidates' reactions, behaviour and solutions to a nonstandard situation like angry customers (Bika, 2018). Online tests and screening games often incorporate interview scheduling modules unless this task is handled by a separate application requiring HR managers' input for evaluating outcomes and decision-making. Moreover, all the above-delineated screening applications collect and analyse applicants' data which are needed for machine learning to improve AI models and for future reference in HRM analytics.

Organisations are employing ICTS extensively for efficient candidate sourcing, screening and interview scheduling. ICTS solutions assist HR managers in finding the best candidates for the vacant position while saving resources and streamlining the recruitment process. However, as ICTS represent a complex ecosystem encompassing the interactions between machines and humans, errors may still require AI solutions to tackle an auxiliary issue of bias in hiring decisions.

2.2 *Mitigating Bias in Hiring Decisions with AI*

Even with recent technological advances, the risk of improper hiring decisions is still high. In this context, 43% of recruiters said they had made errors in hiring mainly due to urgency in filling a vacancy (CareerBuilder, 2017). Whether conscious or unconscious, human factors can badly affect recruiting by leading to improper decisions as well as impeding attempts to maintain diversity and inclusion within businesses. In this regard, AI provides substantial aid to reduce possible bias that may occur in hiring decisions. However, HR managers and ICTS solution developers must follow several precise guidelines, delineated below, to mitigate bias and ensure objectivity and justness in recruitment decisions.

First, it is necessary to develop precise job descriptions and postings. They must use unequivocal language relevant to the open position and understandable to potential applicants. Next, ICTS developers should make provisions for AI solutions in candidate selection to rely on data pertaining to applicant background, work experience and skills instead of non-relevant information. In this sense, it is also significant to maintain anonymisation in applicant screening so that the AI model does not consider personal candidates' demographic data as independent variables. These data must not be accessible to HR managers as well. Third, for robust ICTS functioning, developers must continuously monitor machine learning models and improve them regularly. In addition, to maintain inclusivity in hiring, ICTS developers have to train AI models on various datasets to secure the consideration of candidates with different backgrounds and peculiarities (Köchling & Wehner, 2020). Finally, it is important for ICTS solutions to verify that applicants with various backgrounds and peculiarities are evaluated equally and fairly when the applicants' screening process employs tests, games and one-way interviews (Fig. 2).

Fig. 2 Example of ICTS evaluation of online interview. Source: SHL (2020)

Although sophisticated ICTS solutions can significantly assist in mitigating bias in hiring decisions, they have their restrictions and imperfections. Organisations should consider ICTS as effective but still limited and unable to completely replace human engagement. Thereby, input from HR managers is still essential for proper final decision-making in the recruiting process.

2.3 *Case Studies of Organisations Successfully Using AI in Recruitment*

2.3.1 Brother International Corporation

A leading home office and business equipment supplier, Brother International Corporation, demanded a more vigorous strategy to promote their employer brand and appeal to talents. Driven by this goal, they developed and deployed AI-powered solutions to endorse their brand and continuously attract high-quality professionals. They began by creating a career site to promote their genuine employer brand, offer applicants personalised job suggestions, and house a conversational chatbot that would act as site visitors' initial point of contact with the business. Brother International Corporation updated its CRM system to assist in identifying passive candidates and collecting qualifying leads after the new and enhanced career site went live. They also relied on AI-driven talent analytics that automatically utilised data insights to optimise application alternatives for candidates.

Without this AI platform, the company's recruiters would not be capturing any leads from the chatbot, similar job subscriptions, or partial applications, especially for their more prominent volume roles in sales and marketing. This kind of efficiency is quite helpful, according to Brother's officials. Following the redesign of its career site and the introduction of novel AI-driven recruitment ICTS solutions, the company achieved the following results

- 140% increase in completed job applications;
- 45% web page visits increase;
- 40% job applicant growth;
- 15% increase in candidates who previously worked for Brother;
- 25% reduction in time to fill the open position.

The company has determined that hosting a career site and utilising AI solutions in recruitment is beneficial because these advancements have streamlined the recruitment process and created positive applicant experiences (Blehar, 2023).

2.3.2 AI Recruitment at IBM

Effective recruiting prioritisation predisposes a system to thorough applicant selection at a large company like IBM. The company requires a more effective method of assisting recruiters in seeking the best candidates for available positions and prioritising crucial job requirements. The created solution, IBM Watson Recruitment (IWR), employs AI to make use of job market data and recruiting history to forecast time to fill and select the applicants with the highest likelihood of success (Fig. 3).

AI frees up time for the recruiter to concentrate on the most critical aspects of recruiting: cultivating connections with applicants prioritising and ranking candidate fit. AI creates a match score between a CV and necessary skills based on job

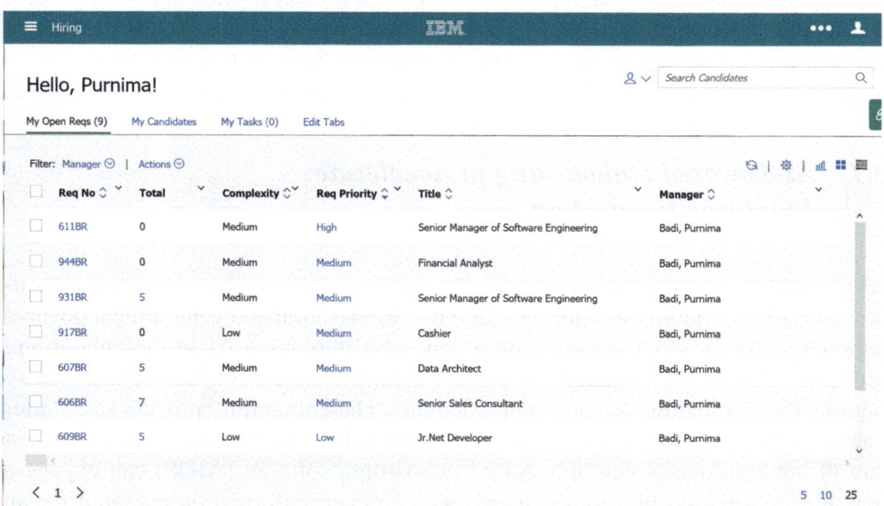

Fig. 3 IBM Watson Recruitment (IWR) screen. Source: IBM (2023)

description. According to biographical information in the CV (such as whether or not they have led a team), the solution can also produce predicted scores for future job success. IWR also monitors recruiting processes to ensure there is no prejudice in candidate selection. Overall, implementing AI in the hiring process enables quicker and more precise hiring and a better applicant and recruiter experience (IBM Corporation, 2021).

3 AI-Driven Onboarding and Training

Onboarding for a newly hired employee is an important and simultaneously challenging procedure for both the employee and the organisation. The employee's early days in his new assignment are critical for the success of his job, and at the same time, the organisation needs to ensure that the open position is appropriately filled. Employee onboarding implies the process of a recruit's induction into a new role and incorporation into the new employer company. This process involves educating employees on companies' corporate rules, policies, schedules, culture and delivering specialised knowledge relevant to execute employees' concrete professional functions in the organisation (Makarius et al., 2020). The purpose of the onboarding process implies fast and efficient employee integration in the company's work routine and ensuring a new hire's contribution to achieving the company's goals. Until recently, the conventional onboarding process commonly required lengthy training sessions, piles of documentation and the involvement of various company employees diverted from carrying out their routine job responsibilities (Fallmann, 2023). Similarly to recruitment procedures, AI solutions in employee onboarding and training are being developed rapidly to address these issues. This section delineates the approaches, specifics and advantages of the use of AI-powered applications in the post-hire processes relevant to employee onboarding and training.

3.1 AI-Powered Onboarding to Accelerate Employee Integration

AI applications rationalise, personalise and improve an employee onboarding process providing smooth incorporation into the organisation and generating a positive experience for an employee starting a new job. Employing AI in the onboarding processes diminishes the need for manual administrative work and delivers many options for automating onboarding procedures. The onboarding process starts after negotiations, job offer acceptance and the completion of formalities before the first day of the employee's new job. An AI onboarding solution (AIOS) can engage a new hire by inviting him to a platform which covers information essential for an effective job start with the company. Such information may be relevant to office

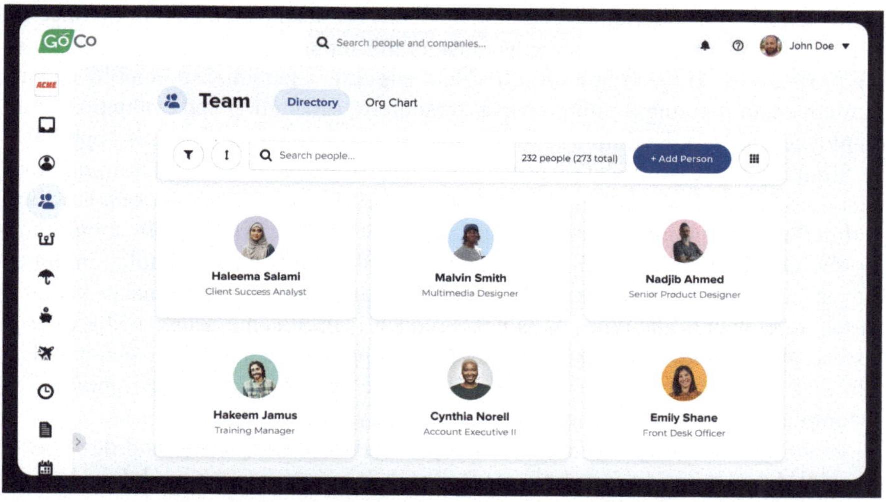

Fig. 4 Example of AI-powered employee onboarding platform. Source: GoCo.Io (2023)

hours, amenities, important contact information, benefits, etc. (Fig. 4). Companies may also pair the AIOS platform with a chatbot to get instant answers to a new employee's questions.

The same AIOS platform expedites the new employee's onboarding process once he has started with a new organisation, helping to familiarise him with the company's regulations, policies, and procedures. On the first day, for instance, AIOS can help employees present their documentation for HR processing through scanning, digitally signing and uploading documents to the platform. Then, during the employee probation period, AIOS regularly monitors and generates reports accompanied by the supervisor's feedback on employee performance and indicates the areas where performance needs improvement. Additionally, well-developed AIOS should support new employee performance by providing them with access to the company's IT resources. Finally, AIOS systems also offer a perfect means for a new employee's social engagement by allowing him to create relationships with colleagues through message boards, online events and interest groups.

3.2 AI's Role in Personalised Learning and Development Programmes

During the process of new hire onboarding, AIOS uses predicted analytics for the development of training programmes. For these purposes, it extracts new employees' data to develop and suggest customised learning and training programmes grounded in their skills, experience, education and functions in the organisation. While sculpting a comprehensive training programme, the employee is given the option to share his individual needs and demands for the development of his specific individual competencies. AIOS may consider this input in the development of training programmes (Fig. 5). As a result, AIOS suggests a personalised employee curriculum encompassing training courses, resources, scheduling and evaluation so an employee can start the training programme once he has his supervisor's approval.

Similarly to the gamification utilised in the recruitment process, AI training modules can also incorporate this concept for more effective engagement of a new hire within the training process (Jarrahi et al., 2023). However, AIOS applications commonly include more sophisticated applications like full business simulation multiplayer games (Fig. 6). In addition, employee training gamification makes use of a variety of elements and types of games. Frequently used game genres include sandboxes, role-playing, and quests. Besides leveraging engagement, these games entertain and thrill trainees creating a more positive and productive environment for accomplishing a training programme successfully.

To upgrade and improve the efficiency of AIOS, AI engineers and developers must integrate an option for employee feedback in the application modules. Quizzes and surveys not only additionally engage employees in their onboarding but also make it feasible to source information on employee experience, perception and evaluation of the process (Fig. 7). The received information aids HR managers in locating grey areas, flaws or stumbling blocks in the AIOS applications and points to the parts and elements that need improvement, optimisation or updating.

In summary, organisations can benefit from AIOS for their HRM areas as these AI solutions speed up employees' integration, improve their experience, and ensure

Fig. 5 Example of AIOS employee customised training programme interface. Source: Heier (2023)

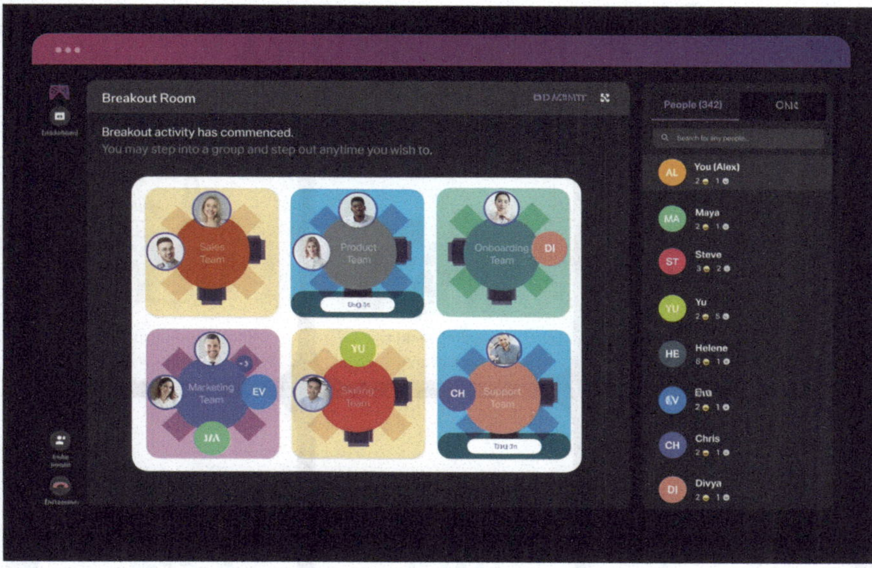

Fig. 6 Example of employee training gamification. Source: Benefits, & Practices (2023)

support received by the new hires throughout the process of onboarding and training. Being delivered by AI-powered onboarding solutions, automation, personalisation, employee engagement, skills, and knowledge upgrade provide a faster return on an organisations' investment in new employees and facilitate their performance from the first weeks of employment.

3.3 *AI-Driven Training Modules in Employee Skill-Building Improvement*

The accomplishment of the training programme by a new employee during the onboarding process does not imply that the process of this development has been completed. Training and development in contemporary organisations are an ongoing process that gives employees many opportunities to upgrade their expertise and obtain new skills. In this context, AI solutions are suitable for identifying gaps in employees' skills through AI-assisted evaluations and providing instant feedback to the employee. Companies may also intermingle AI-assisted evaluation with virtual AI-powered tutor modules, which can coach new hires individually, taking into account the personality of every employee (*Businesstechweekly*, 2023).

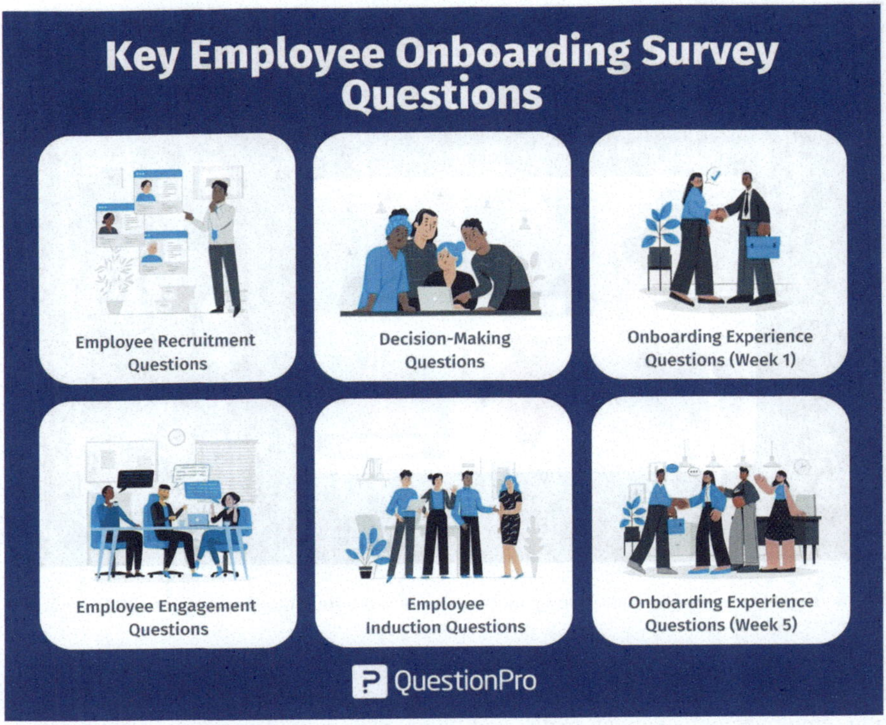

Fig. 7 Example of employee feedback on the onboarding experience survey. Source: Bhat (2018)

4 Conclusions: Embracing AI in HR Recruitment, Training and Onboarding Procedures for Future Success

This chapter has scoped the most recent advancements in AI integration, underpinning several essential HRM procedures. By leveraging a combination of effectiveness and precision, AI has been shown to improve and change recruitment, onboarding, and training practices. That said, AI deployment results in substantial advantages for both businesses and newly hired professionals. We have demonstrated that using ICTS, AIOS, and other AI solutions streamlines employee recruitment, onboarding and training. Implementing AI solutions also indicates the onset for future success in the noted HRM practices. Although adopting AI for HR practices began to become more common not so long ago, it should not be deemed a trendy 'toy' which will be succeeded by more novel technology.

The use of AI in HRM is a fundamental and key paradigm for the development of a company's function, from now and for many years ahead (Kolbjørnsrud et al., 2016). AI has proved an effective HRM instrument in almost all business sectors, from the fast-food industry, e.g., JustEat, to high-tech industries specialised in developing and marketing sophisticated product solutions (Zavvy.Io, 2023). As an organisation's success pivots on employees' skill levels, experience and quality of job performance, it is critical to attract the best talents available in the labour market. Organisations must create and implement efficient AI solutions to gain a competitive edge in attracting and hiring a highly qualified workforce in their endeavours to achieve this success.

Nonetheless, although AI adoption in employee search, recruitment, and onboarding delivers many of the above-explicated benefits for an organisation, the same domain is an area of growing concern for various stakeholders engaged in these HRM processes (Johnson et al., 2022). The most critical disadvantage of AI in HRM comes from the possibility of discriminating against candidates and employees due to undetected bias in data, various errors in the AI algorithms, and machine learning from historical data that reflects pre-existing prejudices of HR managers. Thus, it is paramount to regularly test, validate and improve AI algorithms to make them adequate for an organisation's inclusion and diversity goals. Next, there is a growing concern in the realm of data, candidate evaluation and transparency around AI-driven decisions. To improve stakeholders' trust in AI, this concern can be tackled by creating detailed documentation to accompany the developed AI models with a straightforward elucidation of how the model works and makes its decisions. Finally, AI utilisation in HRM processes is not without ethical concerns in terms of data privacy and security domains. Organisations must develop reliable IT security systems to prevent data leaks, anonymise sensitive data, impose relevant internal policies for data protection and comply with the local state privacy code. Addressing the noted concerns helps mitigate their consequences and facilitates the construction of responsible AI systems in HRM (Malik et al., 2021).

This chapter highlights several implications significant to the theory and literature by delving into the ways in which AI shifts critical human resources management functions in an organisation. First, it systematically analyses how AI can transfigure HRM recruitment processes, ensuring a more efficient and unbiased selection of candidates. Also, the chapter explores the theoretical foundations of AI applications in training and employee onboarding, shedding light on the potential of intelligent systems to improve training outcomes for new hires and companies, as well as to simplify new employees' assimilation into their employing organisation.

In conclusion, it is evident from the business practices which demonstrated excellent results of AI infusion in HR employee recruitment, onboarding and training that AI integration offers unlimited opportunities for HRM. The core advantage of AI deployment is that it leverages HRM from its traditional role of administering the organisation's human resources towards a powerful catalyst fostering better organisational performance (Mikalef & Gupta, 2021). Thereby, it is very likely that organisations will increase their investment in the development of more advanced AI-powered solutions for key HRM processes. In this domain, future AI solutions will widely integrate the latest and trending technologies like robotics, augmented, virtual reality (AR/VR) applications, 3D holographic imaging and others while ensuring compliance with CSR, ethical and legal standards (Dattner et al., 2019). These noted advanced technologies will provide further breakthroughs in candidate interviewing, new employee onboarding and training processes. With the incorporation of the noted technologies in ICTS and AIOS applications and further integration of AI in HRM processes, organisations will thrive and enjoy their future success.

References

Becker, K., & Bish, A. (2021). A framework for understanding the role of unlearning in onboarding. *Human Resource Management Review, 31*(1), 100730. https://doi.org/10.1016/j.hrmr.2019.100730

Benefits, & Practices. (2023). *What is gamification? Benefits and best practices*. https://disprz.ai/blog/gamification-for-employee-learning

Bhat, A. (2018). Employee onboarding: Survey questions + best practices. *QuestionPro*. https://www.questionpro.com/blog/employee-onboarding/

Bika, N. (2018). *Gamification in recruiting: How and why to give it a shot. How to recruit and hire better.* https://resources.workable.com/stories-and-insights/gamification-in-recruiting-effectiveness

Bingham, B., Ilg, S., & Davidson, N. (2002). Great candidates fast: On-line job application and electronic processing: Washington State's new internet application system. *Public Personnel Management, 31*(1), 53–64. https://doi.org/10.1177/009102600203100106

Black, J. S., & Esch, P. (2020). AI-enabled recruiting: What is it and how should a manager use it? *Business Horizons, 63*(2), 215–226. https://doi.org/10.1016/j.bushor.2019.12.001

Blehar, M. (2023). 6 examples of companies successfully using an AI recruiting platform. *Phenom*. https://www.phenom.com/blog/examples-companies-using-ai-recruiting-platform

Businesstechweekly. (2023). Artificial intelligence and its role in employee training and development. *Businesstechweekly.Com*. https://www.businesstechweekly.com/hr-and-recruitment/learning-development/artificial-intelligence-in-training-and-development

Careerbuilder. (2017). 43% of employers have made a bad hire due to lack of (or insufficient) background check. *Careerbuilder.Com*. https://resources.careerbuilder.com/news-research/bad-hire-due-to-insufficient-background-check

Chowdhury, S., Dey, P., Joel-Edgar, S., Bhattacharya, S., Rodriguez-Espindola, O., Abadie, A., & Truong, L. (2023). Unlocking the value of artificial intelligence in human resource management through AI capability framework. *Human Resource Management Review, 33*(1), 100899. https://doi.org/10.1016/j.hrmr.2022.100899

Consulting, C. I. T. (2023). *ATS Consulting Services-CrossTab IT Consulting*. https://crosstab.in/ats-consulting-services/

Corporation, I. B. M. (2021). The business case for AI in HR. *IBM*. https://www.ibm.com/downloads/cas/A5YLEPBR

Corporation, I. B. M. (2023). AI in recruitment-IBM Blog. *IBM*. https://admin01.prod.blogs.cis.ibm.net/blog/ai-in-recruitment/

Dattner, B., Chamorro-Premuzic, T., Buchband, R., & Schettler, L. (2019). The legal and ethical implications of using AI in hiring. *Harvard Business Review, 25*, 1–7.

Davila, N., & Pina-Ramirez, W. (2018). *Effective onboarding*. American Society for Training and Development.

Esch, P., Black, J. S., & Ferolie, J. (2019). Marketing AI recruitment: The next phase in job application and selection. *Computers in Human Behavior, 90*, 215–222. https://doi.org/10.1016/j.chb.2018.09.009

Fallmann, D. (2023). *Council Post: How can AI-powered search accelerate employee onboarding*. https://www.forbes.com/sites/forbestechcouncil/2023/07/19/how-can-ai-powered-search-accelerate-employee-onboarding/

GoCo.Io. (2023). All-in-one HR software-online HR, payroll & benefits. *GoCo.Io*. https://www.goco.io/

Gomes, O. (2023). *Guide to resume screening*. https://blog.talview.com/how-to-screen-resumes-like-a-pro

Heier, Ø. (2023). Snapper and GO1 enter into a collaboration agreement to offer comprehensive e-learning courses to Snapper's customers-Snapper. *Snapper*. https://snapper.no/en/snapper-and-go1-enter-into-a-collaboration-agreement-to-offer-comprehensive-e-learning-courses-to-snappers-customers/

Jarrahi, M. H., Möhlmann, M., & Lee, M. K. (2023). Algorithmic management: The role of AI in managing workforces. *MIT Sloan Management Review, 64*(3), 1–5.

Jobvite. (2021). What is a recruiting Chatbot. *Jobvite*. https://www.jobvite.com/glossary/recruiting-chatbot/

Johnson, B., Coggburn, J., & Llorens, J. (2022). Artificial intelligence and public human resource management: Questions for research and practice. *Public Personnel Management, 51*, 538–562. https://doi.org/10.1177/00910260221126498

Johnson, R. D., Stone, D. L., & Lukaszewski, K. M. (2020). The benefits of eHRM and AI for talent acquisition. *Journal of Tourism Futures, 7*(1), 40–52. https://doi.org/10.1108/JTF-02-2020-0013

Kaplan, A., & Haenlein, M. (2019). Siri, Siri, in my hand: Who's the fairest in the land? On the interpretations, illustrations, and implications of artificial intelligence. *Business Horizons, 62*(1), 15–25. https://doi.org/10.1016/j.bushor.2018.08.004

Köchling, A., & Wehner, M. C. (2020). Discriminated by an algorithm: A systematic review of discrimination and fairness by algorithmic decision-making in the context of HR recruitment and HR development. *Business Research, 13*(3), 795–848. https://doi.org/10.1007/s40685-020-00134-w

Koechling, A., Wehner, M. C., & Warkocz, J. (2023). Can I show my skills? Affective responses to artificial intelligence in the recruitment process. *Review of Managerial Science, 17*(6), 2109–2138. https://doi.org/10.1007/s11846-021-00514-4

Kolbjørnsrud, V., Amico, R., & Thomas, R. J. (2016). How artificial intelligence will redefine management. *Harvard Business Review, 2*(1), 3–10.

Makarius, E. E., Mukherjee, D., Fox, J. D., & Fox, A. K. (2020). Rising with the machines: A sociotechnical framework for bringing artificial intelligence into the organization. *Journal of Business Research, 120*, 262–273. https://doi.org/10.1016/j.jbusres.2020.07.045

Malik, N., Tripathi, S., Kar, A., & Gupta, S. (2021). Impact of artificial intelligence on employees working in industry 4.0 led organizations. *International Journal of Manpower*. https://doi.org/10.1108/ijm-03-2021-0173

Mikalef, P., & Gupta, M. (2021). Artificial intelligence capability: Conceptualization, measurement calibration, and empirical study on its impact on organizational creativity and firm performance. *Information & Management, 58*(3), 103434. https://doi.org/10.1016/j.im.2021.103434

Muse, T. (2016). What is a digital interview and how does it work? *The Muse*. https://www.themuse.com/advice/how-to-nail-your-video-assessment-and-actually-get-to-meet-the-hiring-manager-in-person

Pan, Y., & Froese, F. J. (2023). An interdisciplinary review of AI and HRM: Challenges and future directions. *Human Resource Management Review, 33*(1), 100924. https://doi.org/10.1016/j.hrmr.2022.100924

SHL. (2020). SHL smart interview to integrate with Microsoft Teams and Zoom. *AiThority*. https://aithority.com/video/shl-smart-interview-to-integrate-with-microsoft-teams-and-zoom/

Weitzman, T. (2022). The top five ways AI is transforming business. *Forbes.Com*.

Zavvy.Io. (2023). *AI in HR examples: 9 companies successfully using AI innovation in key people processes*. https://www.zavvy.io/blog/ai-in-hr-examples

The Behavioural Science of Using AI in HRM Decision-Making: When It Helps and When It Goes Wrong

Craig Graham Anderson

Abstract This chapter demonstrates how behavioural science can be used in identifying opportunities and costs of applying Artificial Intelligence (AI) in human resource management (HRM). The digitisation of HRM has examples of behavioural biases informing their design. Similarly, the use of AI, as an extension of the digitisation of HRM can also have heuristic rule-based decision-making determining their design. At the same time, behavioural science has also been used to support HRM in the context of behavioural biases, cognitive biases, and heuristic rule-based decision-making. With the synergy between cognitive bias, and heuristic rule-based decision-making, behavioural science has been used to consider the opportunities and costs in applying AI. This chapter combines the literature on AI and HRM, as well as behavioural science and AI, to provide a novel tool for considering the opportunities and costs of applying AI in HRM. The analysis at the intersection of behavioural science, AI, and HRM is gives illustrative examples and debates the opportunities and costs of applying AI in HRM from the perspective of behavioural science. The examples given can be used for further application of behavioural science as a tool for considering the opportunities and costs in applying AI in HRM.

1 Introduction

Digital tools to manage human resource management (HRM) have expanded significantly. To date, the application of behavioural science to HRM has largely focussed on using behavioural insights to inform the creation of digital tools that help de-bias recruitment decision-making. However, there are fewer applications that look at how to use behavioural science to understand when a digital tool is needed. This chapter seeks to show how behavioural science can be used to inform

C. G. Anderson (✉)
Stirling Management School, University of Stirling, Stirling, UK
e-mail: c.g.anderson@stir.ac.uk

M. T. Del Val Núñez et al. (eds.), *Artificial Intelligence and Business Transformation*, Contributions to Management Science,
https://doi.org/10.1007/978-3-031-58704-7_14

decisions in a digitised recruitment world. The role of artificial intelligence (AI) in HRM and when to use it has been considered (Black & van Esch, 2020). However, this is situated in practical implications and processing large amounts of data. The contribution of this chapter will be to reflect on the position on behavioural science in relation to this. The methodology applied in this chapter could be applied to the wider operational areas of HRM. The chapter can act as a tool for a behavioural science lens to applied in other areas and HRM and organisational behaviour.

Before proceeding it is important to classify some useful definitions in behavioural science, as well as the use of digital tools and AI in human resource management as these phenomena are used alongside each other or aggregated under 'digital HRM' (Meijerink et al., 2021):

Heuristics: Used to reduce the cognitive capacity needed to consider a given problem or choice; calculations are based on incomplete information instead (Groner et al., 2014). For example, assuming that objects seen with less clarity are further away, but clarity can be determined by visibility, so clarity is not always consistent with distance (Tversky & Kahneman, 1974).

Cognitive bias: Relates to where judgments are made intuitively, and that intuition is guided by perception. People can have limited willpower; they can be tempted and can be short- sighted (Jolls et al., 1998). For example, in a changing environment it can be difficult to judge the likelihood and value of present and future events. A further example is that we care what others think, as well as about our own identity, making it difficult for us to place accurate valuations on others.

Digital tools in human resource management: Can be inclusive of e-HRM, online HRM, and digital HRM. Broadly this refers to moving physical or manual processes to digital platforms, providing structure, streamlining, and efficiency in decision-making (Malik et al., 2020).

Artificial intelligence (AI) in human resource management: In AI in HRM databases are merged with language models and imaging tools to perform tasks and make decisions instead of humans through machine learning and knowledge-based systems (Malik et al., 2020).

Workforce analytics: The focus on, and use of, metrics to drive and inform strategic decision-making in talent management, including predictive analytics through big data defined in terms of the volume, variety, velocity, variability, veracity, and complexity of information collected (Huselid, 2018).

2 The Use and Emergence of Digital Tools in HRM

It has been known for some time that decision-making in organisations, including HRM, can be based on heuristics and cognitive bias. In addition, wider social biases are managed through HRM, such as diversity. Digital or e-HRM tools emerged to help design recruitment in a way that would lead to less bias. Some of these assisted in blinding biasing information or emphasising other information. They also helped in designing recruitment processes which would lead to fewer cognitive biases and heuristic errors (Black & van Esch, 2020). This also drew behavioural science into the area of HRM, acknowledging the role of social bias, but also wider cognitive biases and heuristics which underpin HRM decision-making. As experiments and new tools emerged across HRM systems there became a suite of digital tools to utilise. This led to the need for, and opportunity to, integrate across digital tools and e-HRM systems. With increasing integration and data there is scope for complex workforce analytics, emphasising large data and analytics. Once patterns can be identified across data, systems, and outcomes, decisions can become more automated and AI tools emerge (Yang et al., 2023).

2.1 The Emergence of AI in HRM

Digital tools have been experimented on in HRM and recruitment for some time. Lawler and Elliot (1993) tested an experimental 'AI' tool to help sort job classifications, acknowledging the the typically heuristic based decision-making. While classifying this tool as an 'AI', the 'expert system' designed to improve human decision-making would be better classified as a digital tool today. This demonstrates not only the need to disentangle interrelated concepts, but also acknowledges the importance of cognitive biases and heuristics in the innovation of digital tools for HRM.

AI is considered to be part of 'digital recruitment 3.0' (Black & van Esch, 2020), to help process information in an increasingly complex recruitment environment. It can be a challenge to separate the use of digital tools, the emergence of 'big data', and the use of AI. In a sense, these different aspects converged to create the use of AI in HRM. As the use of digital tools expanded, increasing data availability and complexity, opportunity emerges to utilise, manage, and process that data in new ways. The collection of raw data in of itself however is not an endpoint. When trying to extract knowledge from data, new rule-based models can emerge. Again, these rule-based models can at first be learned from the heuristics used by decision-makers. With sufficient data, these rules can eventually be followed without human decision-making (Malik et al., 2020).

The emergence and use of AI could limit the capacities of existing theories of organisational behavior, owing to increased use of AI which learns from itself. There is now a non-human actor driving change (van Rijmenam & Logue, 2021), as well as the potential for an Automation-Augmentation Paradox (Raisch & Krakowski, 2021). Automation and Augmentation has the potential to transform most industrial sectors with the potential for widespread socio-economic impacts (Dwivedi et al., 2021). Bankins et al. (2023) identify five implications in the use of AI for organisational behaviour. Human–AI collaboration; perceptions of algorithm and human capabilities; worker attitudes towards AI; AI as a control mechanism in algorithmic management of platform-based work; labour market implications.

2.2 Conceptualisation of the Application of AI in HRM

Böhmer and Schinnenburg (2023) conceptualize the reasons for applying AI in HRM, highlighting pros and cons. These are summarised as resolving transparency ambiguity, data ambiguity, performance ambiguity, and job design ambiguity (Fig. 1).

Malik et al. (2022) look at how and why AI would be applied in strategic HRM with possible application of AI from HRM planning and work design, to recruitment, training, as well as employee performance and benefits. Feedback loops exist between the competencies in applying machine learning and AI, as well as supervision and ethics within its application. This interacts with workplace behaviours resulting in both positive and negative outcomes. The authors recommend interventions across feedback loops and investing in specific areas and personnel where AI can best assist. For example, feedback loops between supervised and unsupervised learning within an AI augmented environment. Options exist to work between the benefits of customisation personalisation, better data in recruiting HRM personnel, as well as training for HRM personnel in AI. Radonjić et al. (2022) look at why and how big data maturity models (BDMM) might develop and be applied in the organisation. There is a feedback loop between increased data extraction and exploitation, feeding into strategic decision-making. More data leads to new information to make decisions upon, however data must be valid and well managed. Organisational willingness has implications for governance of the implication of AI, but new and improved data can also be used to justify decision-making. Ultimately careful alignment and upskilling is required.

Transparency Ambiguity:

- Pro: Better information and participation. Access to information leading to empowerment.
- Con: Black box of algorithms leading to reduced understanding of data and decisions

Data Ambiguity:

- Pro: Big data fundamental to new conclusions through structuring of data
- Con: Legal and ethical conditions of data collection and handling

Performance Ambiguity:

- Pro: More visibility and objectivity in job performance and better person-job fit.
- Con: Excessive observation, monitoring, and control through AI tools.

Job Design Ambiguity:

- Pro: Enhanced and augmented abilities through AI supported tasks
- Con: Intellectual impoverishment and reduced stimulation

Fig. 1 Summary of Böhmer and Schinnenburg (2023). Source: Böhmer and Schinnenburg (2023)

2.3 *Limitations in the Application of AI in HRM*

The first problem that arises from the use of AI in HRM and big data in the workplace is shifting to using data for real time strategic decision-making as well as shifting away from qualitative assessment. This requires a new set of skills for HRM managers. Deciding on the scope of data collection raises practical and ethical concerns (Zhang et al., 2021). Processes need to be aligned between recruitment objectives, the development of strategy, and recruitment activities. Strategy and objective setting that is still dominated by human decision-making that might not be aligned to the outcomes of and AI driven process. There is a need to consider the human-AI configuration including employee to customer, peer to peer, and manager to team interactions (Budhwar et al., 2022).

One of the risks of in the use of AI in HRM is the need for oversight and auditing of its use. Prikshat et al. (2022) consider a multi stakeholder approach to managing the use of AI in HRM in terms of fairness, accuracy, impact, intersectionality, explainability, responsible design, predictability, reliability and viability, ownership, consent and confidentiality, and data protection. Each of these aspects would have different stakeholders implicated in managing and receiving the impacts of these. The level of oversight required to ethically apply AI in HRM is a considerable resource consideration. A lack of human interface in an inherently human process of human resource management can be a common criticism of the use of AI as well as e-HRM more widely, reducing personal interaction and increasing distance between the applicant or employee and organisation (Malik et al., 2020). These questions of human-AI interaction can be investigated from a behavioural science lens.

3 The Use and Emergence of Behavioural Science in HRM and Recruitment

As depicted throughout this chapter the role of heuristics and cognitive biases in HRM decision-making has underpinned some digital tools and AI platforms that exist in HRM. However, formal integration of behavioural science into HRM can be a difficult to trace. This is partially owing to the variety of job roles and disciplines that can interact with organisational behaviour, psychology, and HRM. Behavioural science can span economic, social, systemic, and psychologically driven disciplines and roles. By the 1970s HRM had become a psychology driven practice, while labour and industrial economics remained largely separate (Kaufman, 2014). However, the emergence of strategic HRM, as well as the input of psychology into economics, including personnel economics, led to a convergence where behavioural economists began tackling problems alongside HRM researchers (Fig. 2).

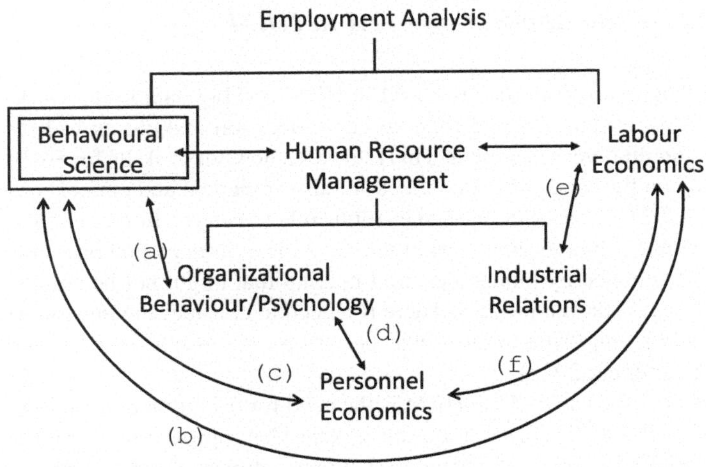

Interaction (a): Insights from organizational behaviour are incorporated behavioural science and vice versa. Social comparison bias, the tendency, when making hiring decisions, to favour potential candidates who don't compete with one's own strengths (Garcia, Song & Tesser, 2010), being one of few examples. Concepts could be integrated more (Backes-Gellner et. al., 2008; Lepak &Shaw, 2008; Lazear & Shaw, 2007).	Interaction (c): Where insights from behavioural science are incorporated into personnel economics. It is in this interaction that the sub-discipline of micro-foundations lies, which is a prominent example of adopting the influence of behavioural science into the study of employment.	Interaction (e): The interaction between industrial relations and labour economics. In terms of references to HRM and labour economics, this interaction has the most interdisciplinary citations.
Interaction (b): Where the insights from behavioural science are incorporated into labour economics. Interaction (b) is an active bridge between sub-disciplines, with behavioural science concepts being applied to pensions (Thaler & Bernartzi, 2004), retirement (Bidewell, Griffin & Hesketh, 2006), job search (Paserman, 2008), wages (Schoenfelder & Hantula, 2003), bonuses (Hesketh, 2000; Shelley & Omer, 1996).	Interaction (d): Underexploited given the work of micro-foundations and increasing integration of behavioural science into personnel economics. Limited interaction between personnel economics, HRM, or organizational behaviour (Barney & Felin, 2013; Lazear & Shaw, 2007; Lepak &Shaw, 2008).	Interaction (f): The most active disciplinary bridge. Personnel economics is a subfield of labour economics, and they share the same approaches; albeit they have different focuses and organizational scales.

Fig. 2 Behavioural science in employment research. Source: Anderson (2017)

4 Behavioural Science and AI

Bickley and Torgler (2023) bring together a range of behavioural models for how behavioural science interacts with AI. The first considers the scale of the effect and informational complexity. Traditional computing of linear models is small effect and low informational complexity. Classical AI is of moderate effect and complexity and is driven by qualitative heuristic-based logics. High complexity environments can have high impact with more specific and case-based logics. With different levels of effect and complexity there is trade-off between the intractability of a complex model, the drive for rational outcomes, and the need for outcomes to reflect the behaviours of humans. Complexity and decision-making develop over time. Systems need to consider long and short-term feedback loops.

Balasubramanian (2021) highlights that human decisions are driven by dispositional factors, heuristics, and limited information. AI decisions are driven by data, algorithm complexity, and the ability to share, store and process data. When considering the interface between these two decision mechanisms, speed, consistency, risk, and efficiency need to be considered. Iteration is required to make decisions

Mills, Costa, & Sunstein (2023): The Opportunities and Costs of AI in Behavioral Science

Opportunity 1: Identifying Biases – Predictable patterns in AI data
Opportunity 2: Integrating Heterogeneity – Individual data with feedback loops
Opportunity 3: Handling Complexity – Systems thinking and utilizing 'leverage points'

Costs: Data handling, implementation, experiential, trust, transparency, skills, information asymmetry, maladaptive responses - Need to have a 'human in the loop'.

Fig. 3 Summary of Mills et al. (2023): Source: Mills et al. (2023)

and systems robust. Both humans and machines make judgments to support institutional decision-making. Decisions have consequences on recipients, so it is important to understand the link between AI inference and AI behaviour (Pedersen & Johansen, 2020).

Mills et al. (2023) consider the intersection between behavioural science and AI decision-making. Firstly, it is noted that behavioural science identifies predictable patterns of behaviour, with that predictability it is likely that AI will be able to identify biases from data. Secondly, AI has the capacity to integrate multiple variables allowing for iterative models which have individual feedback loops rather than the linear models of more simple behavioural models. Thirdly AI has the capacity to situate behaviours within a wider complex system, there is opportunity to utilise AI to identify where behaviours have the greatest interaction within that system. They also identify costs similar to those listed in the limitations in the application of AI section of this chapter (Fig. 3).

Mills et al. (2023) acknowledge the costs imposed by AI, including the need to have a human in the loop. There is a need for human interaction in the use of AI and HRM, given increased distance between the organisation and applicant or employee (Malik et al., 2020).

5 Behavioural Science of Using AI in HRM Decision-Making

In the following three subsections of this chapter, a model has been developed to consider the strengths and weaknesses of the application of AI across HRM decision-making. The examples given are only illustrative. The premise is to provide a framework for considering why the use of AI may help decision-making based on the behavioural biases that may be underpinning the decision-making context. Importantly it explores the opportunities and costs in choosing to implement AI.

5.1 *Opportunity 1: Identifying Biases—Predictable Patterns in AI Data*

Identifying biases through predicable patterns in data was the first opportunity identified by Mills et al. (2023) in the intersection between AI in behavioral science. The following four example scenarios identify areas where ambiguity and cognitive biases or heuristics exist within HRM decision-making (Fig. 4).

Transparency Ambiguity Opportunities and Costs: In CV/Resume parsing/sifting, AI provides an opportunity to identify where bias is occurring, with increased information and data on which factors on resumes are resulting in preferable selection. Research has suggested that, when confronted with a pile of job applications, recruiters follow a strategy of picking applicants with positive characteristics ('diamonds') rather than eliminating applicants with negative characteristics ('lemons') (Eriksson & Rooth, 2014). With better data and AI there is an opportunity to identify which positive and negative aspects are resulting in selection. There can also be better matching to essential and desired criteria and better communication to applications to increase transparency in allowing them to understand what is prioritised in selection. However, with more candidates being successful at matching their application to the desired criteria of the job, it becomes much harder to differentiate between increasingly similar applications. This could lead to distinction bias where the magnitude of small differences can seem greater when comparing side by side than their actual real-world difference (Hsee & Zhang, 2004). Also, it is likely that applicants themselves will use AI to craft their application, with increasing accuracy and improved matching in essential and desired criteria.

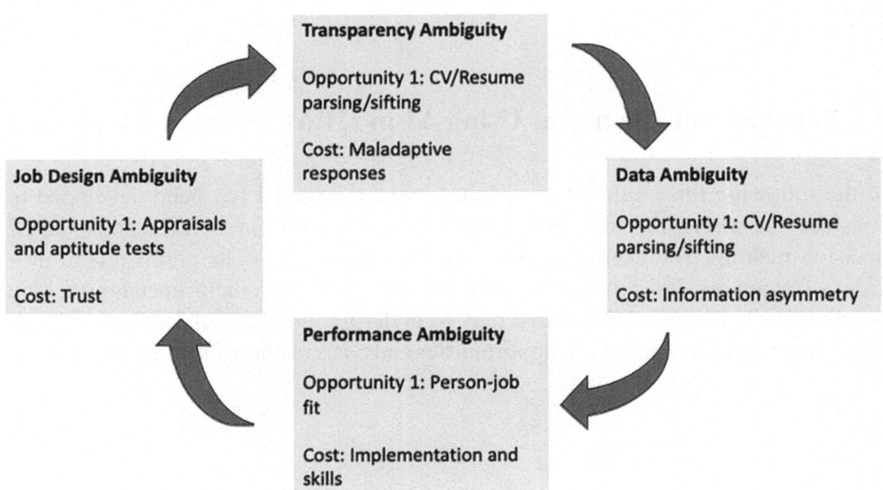

Fig. 4 Opportunity 1. Source: Author. Adapted from Böhmer and Schinnenburg (2023); Mills et al. (2023); Malik et al. (2022); Radonjić et al. (2022)

Data Ambiguity Opportunities and Costs: The essential and desired criteria of an advertised position, collectively termed selection criteria, also represent a possible scenario whereby asymmetry of information within the recruitment process can cause adverse selections (Akerlof, 2002). Asymmetry of information arises when not all stakeholders have equal relevant information (Rasmusen, 2001). Data ambiguity is reduced by better structuring of data. The link between a CV/resume being successful at the parsing/sifting phase and leading to successful performance in subsequent recruitment activities and job performance can be identified. Importantly, detail in this data could identify specific traits within a CV/resume that are paired with other traits subsequently learned about the candidate to understand better the link between CV/resume parsing/sifting and person-job or person-organisation fit. AI could provide a significant contribution in being able to identify and communicate at an early stage where potential performance predictors can be identified from the CV/resume. Successful applicants who join the organisation only know they were successful; they do not know their weaknesses. Complexity and decision-making also develops over time and more complex systems need to consider long and short-term feedback loops (Bickley & Torgler, 2023). However, using this data requires the skills and systems to manage potential performance risks.

Performance Ambiguity Opportunities and Costs: Typically, *person-organization fit* is measured through congruence of personality and traits of the applicant and the organization. *Person-job fit* is measured by assessing existing skills and experience against the demands of the job. The desire for person-job or person-organization congruence can differ depending on the length of contract tenure (Sekiguchi & Huber, 2011), subjecting it to potential time bias such as 'hyperbolic discounting', where a greater emphasis may be placed on immediate needs than is optimal. AI could be used to simulate potential outcomes and lengths of tenures for a given candidate, including probabilistic outcomes. This could result in a better matching of person-job vs-person organisation fit when recruiting. However, decision makers would require training on how to interpret these simulations.

Job Design Ambiguity Opportunities and Costs: Confirmation bias is a tendency to focus on information that enforces one's own preconceptions (Nickerson, 1998). Egocentric/self-serving bias is a tendency to attribute positive events to one's own character but attribute negative events to external factors (Miller & Ross, 1975) The use of job testing can result in 'confirmation bias' and 'self-serving bias' in response to the test (Ployhart & Harold, 2004; Ryan & Ployhart, 2000), where an applicant will blame a bad score on the test rather than themselves. Where performance is augmented by AI, there may be a tendency to blame the design of the AI over their own performance. So, while AI gives the opportunity to collect and provide more data on performance, there are trust elements in how employees will respond. These are not just issues in excessive observation, but also how employees might respond to positive and negative outcomes. Self-serving bias is not equal across outcomes.

5.2 *Opportunity 2: Integrating Heterogeneity—Individual Data with Feedback Loops*

The second opportunity through the use of AI identified by Mills et al. (2023) is individual data with feedback loops. Importantly this can lead to customised information that would be difficult to provide without the use of AI (Fig. 5).

Transparency Ambiguity Opportunities and Costs: Unsuccessful job applicants can be frustrated about the lack of transparency in their search for a position, and a lack of feedback about the flaws of their profiles (Martinez-Gil, 2014). While AI has the capacity for personalisation, there arises a transparency ambiguity where the recipients of that personalised feedback might not know whether that feedback has been AI generated or not. They do not know why certain elements have been emphasised in this feedback. AI decisions are driven by data, algorithm complexity, and the ability to share, store and process data. Iteration is required to make decisions and systems robust; speed, consistency, risk, and efficiency need to be considered (Balasubramanian, 2021).

Data Ambiguity Opportunities and Costs: Given the rise of the internet and online applications, human resource managers have to select from a greater pool of candidates. However, traditional economic theory suggests that the supplier (candidate) knows the true quality of him/herself. Asymmetry of information, for example in the quality of the labour force, makes it harder for a candidate to know the value of his or her own 'quality' in relation to the specific job. Also, HR managers are unable to estimate the true quality of a specific candidate, given the excess and asymmetry of information. There can be too much contrasting information to process fully and completely. AI could provide the opportunity to track individual

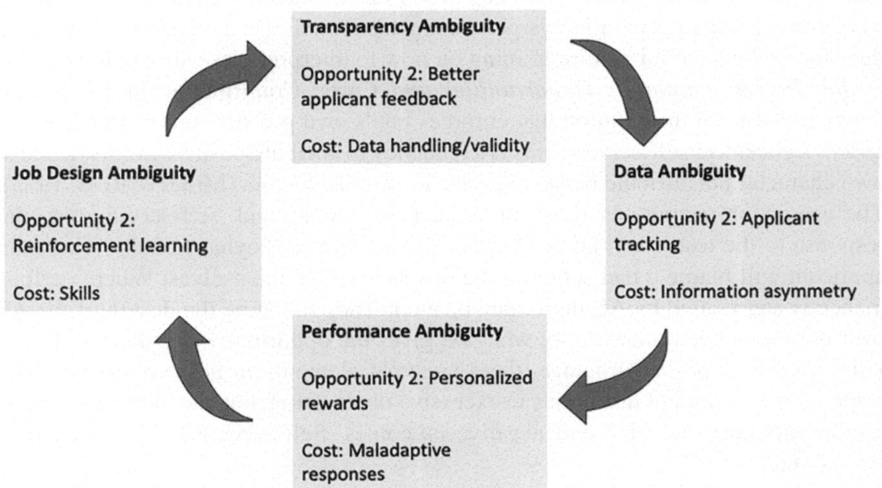

Fig. 5 Opportunity 2. Source: Author. Adapted from Böhmer and Schinnenburg (2023); Mills et al. (2023); Malik et al. (2022); Radonjić et al. (2022)

journeys through recruitment. It can use this data to model and identify which kinds of candidates are not applying or dropping out of the application process. Recruiters do not always have information on why applicants did not apply.

Performance Ambiguity Opportunities and Costs: The sense of responsibility for success can impact the level of satisfaction with rewards, with a sense of low responsibility for success making rewards feel less deserved (Chow & Lowery, 2010). Increased data and personalisation through AI can improve the link between rewards and action. However, rewards can lead to in-groups and outgroups (Kehoe & Wright, 2013). Self-serving bias could lead to those receiving awards feeling a sense of ownership, but those who do not get awards could attribute their lack of award to the AI system.

Job Design Ambiguity Opportunities and Costs: One of the potential strengths of the use of AI in HRM is its potential to track job performance and customise training plans. Importantly complexity and decision-making also develops over time and more complex systems need to consider long and short-term feedback loops (Bickley & Torgler, 2023). Career counselling and the provision of skills assessments would require new skills in processing and advising on data. Short-term skills needs and long-term development might not be linked in a linear fashion. Understanding short- and long-term trajectories will require iterations in the development of the data as well as the advice given.

5.3 Opportunity 3: Handling Complexity—Systems Thinking and Utilizing 'Leverage Points'

The third opportunity identified by Mills et al. (2023) is the capacity for AI to be able to identify the interface between systems and behaviour, allowing for new ways of strategic intervention and strategic alignment (Fig. 6).

Transparency Ambiguity Opportunities and Costs: Goal and target setting is often used as a means by which to achieve strategic objectives. However, individual responses to goal setting can be nuanced. AI gives potential to utilise more data in the setting of targets and goals. In addition, more data can be collected on the nuances of how or why the goals set are met or not met, or successful or unsuccessful. A greater number of metrics of the strategic success of goal and target setting can be collated. However, it is likely that only a limited number of people within an organisation will have access to this data, potentially creating an asymmetry of information or a lack of trust in goals set. The use of AI in HRM is the needs oversight and auditing of its use. Prikshat et al. (2022) consider a multi stakeholder approach to managing the use, but this is potentially costly to the organisation.

Data Ambiguity Opportunities and Costs: Critical incidents are often used to identify and manage risk within an organisation. With more data and the use of AI identifying critical points of interaction, these could be used in strategic decision-making. The use of critical incidents can influence trust and relationships with

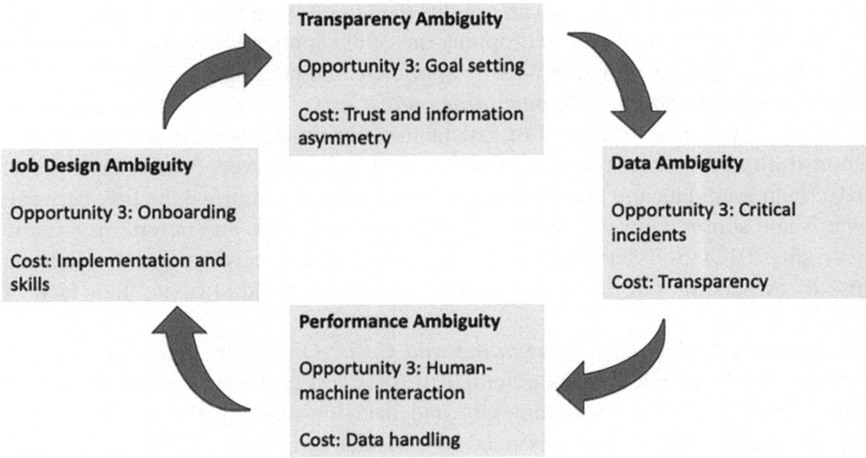

Fig. 6 Opportunity 3. Source: Author. Adapted from Böhmer and Schinnenburg (2023); Mills et al. (2023); Malik et al. (2022); Radonjić et al. (2022)

managers (Krosgaard et al., 2002). As a source of a critical incident, you do not know who else has been a recipient of this or how often it occurs, leading to a base rate fallacy, a failure to adequately consider usual occurrence (base rate) regardless of specific situation (Bar-Hillel, 1980). With increasing data to identify critical incidents, the amount of reporting of incidents may change. Employees would need to be given some indication on the occurrence of these.

Performance Ambiguity Opportunities and Costs: AI provides an opportunity to augment and improve job performance. However, there is a need to monitor the feedback loops between supervised and unsupervised learning within and AI augmented environment. Strategic options exist to work between the benefits of customisation personalisation, better data in recruiting HRM personnel, as well as training for AI amongst HRM personnel (Malik et al., 2022). Data handling needs consideration of where the human action or machine action is leading to outcomes. This may require new analysis and data processing techniques, as well as additional performance measurement.

Job Design Ambiguity Opportunities and Costs: Onboarding is a critical time in an employee's journey. As aforementioned, critical strengths and weaknesses could be predicted through AI. Enabling successful onboarding is a crucial point in an employee's performance and satisfaction. With the use of AI, it may be possible to identify across new entrants, key predictions of job performance and satisfaction stemming from differences in onboarding. However, successful onboarding requires resource to support the new entrants and skills in interpreting and actioning new data feedback loops.

Mills et al. (2023) highlight the importance of 'leverage points' in the intersection of behavioural science and AI. Previous work has highlighted what to look out

AI in HRM	Behavioural Science
Organisational Behaviour: • Data and feedback on CV/resume parsing/sifting • Data on employee performance and skills • Customisation of feedback and rewards	Organisational Behaviour: • Better information congruence and reduced information asymmetry • Trust, in-groups and self-serving biases
HRM Managers: • Skills and training for data and AI	HRM Managers: • Time discounting, short-and long-term strategic aims

Fig. 7 Chapter highlights. Source: Author

for when considering the intersection of 'leverage points' and behavioural science (Black et al. 2023).

6 Chapter Summary

This chapter has sought to present how behavioural science can be used to interrogate the effectiveness of AI in HRM. The digitisation of HRM has roots in behavioural science, cognitive bias and heuristics. Similarly, AI tools can be based on heuristic and rule-based decision-making. Previous work has considered how there are both opportunities and costs in applying AI in HRM. Behavioural science has also been used as a lens for interrogating the application of AI. This chapter therefore demonstrates that behavioural science can be a useful tool for considering the implications of applying AI in HRM. The examples presented could be used as a tool for further interrogation. The following table presents the highlights from some of the examples listed in this chapter (Fig. 7).

References

Akerlof, G. A. (2002). Behavioral macroeconomics and macroeconomic behavior. *American Economic Review, 92*(3), 411–433.

Anderson, C. G. (2017). *Framing employment research using behavioural science*. [Doctoral Thesis, University of Stirling]. Stirling Online Research Repository. http://hdl.handle.net/1893/26373

Backes-Gellner, U., Bessey, D., Pull, K., & Tuor Sartore, S. N. (2008). What behavioural economics teaches personnel economics. *University of Zurich Institute for Strategy and Business Economies Paper,* (77).

Balasubramanian, G. (2021). When artificial intelligence meets behavioural economics. *NHRD Network Journal, 14*(2), 216–277.

Bankins, S., Ocampo, A. C., Marrone, M., Restubog, S. L. D., & Woo, S. E. (2023). A multilevel review of artificial intelligence in organizations: Implications for organizational behavior research and practice. *Journal of Organizational Behavior*. https://doi.org/10.1002/job.2735

Bar-Hillel, M. (1980). The base-rate fallacy in probability judgments. *Acta Psychologica, 44*(3), 211–233.

Barney, J., & Felin, T. (2013). What are microfoundations? *The Academy of Management Perspectives, 27*(2), 138–155.
Bickley, S. J., & Torgler, B. (2023). 3. Behavioural economics: What have we missed? Exploring 'classical' behavioural economics roots in AI, cognitive psychology and complexity theory. In M. Altman (Ed.), *Handbook of research methods in behavioural economics* (pp. 32–59). Edward Elgar Publishing, chapter 3.
Bidewell, J., Griffin, B., & Hesketh, B. (2006). Timing of retirement: Including a delay discounting perspective in retirement models. *Journal of Vocational Behavior, 68*(2), 368–387.
Black, I., Leventon, J., & Anderson, C. (2023). 15. Nudging leverage points: Influencing transformative policy change. In M. Carrigan, V. K. Wells, & K. A. Papadas (Eds.), *Research handbook on ethical consumption* (pp. 251–269). Edward Elgar Publishing, chapter 15.
Black, J. S., & van Esch, P. (2020). AI-enabled recruiting: What is it and how should a manager use it? *Business Horizons, 63*(2), 215–226.
Böhmer, N., & Schinnenburg, H. (2023). Critical exploration of AI-driven HRM to build up organizational capabilities. *Employee Relations, 45*(5), 1057–1082.
Budhwar, P., Malik, A., De Silva, M. T., & Thevisuthan, P. (2022). Artificial intelligence–challenges and opportunities for international HRM: A review and research agenda. *The International Journal of Human Resource Management, 33*(6), 1065–1097.
Chow, R. M., & Lowery, B. S. (2010). Thanks, but no thanks: The role of personal responsibility in the experience of gratitude. *Journal of Experimental Social Psychology, 46*(3), 487–493.
Dwivedi, Y. K., Hughes, L., Ismagilova, E., Aarts, G., Coombs, C., Crick, T., et al. (2021). Artificial intelligence (AI): Multidisciplinary perspectives on emerging challenges, opportunities, and agenda for research, practice and policy. *International Journal of Information Management, 57*, 101994.
Eriksson, S., & Rooth, D. O. (2014). Do employers use unemployment as a sorting criterion when hiring? Evidence from a field experiment. *The American Economic Review, 104*(3), 1014–1039.
Garcia, S. M., Song, H., & Tesser, A. (2010). Tainted recommendations: The social comparison bias. *Organizational Behavior and Human Decision Processes, 113*(2), 7–101.
Groner, R., Groner, M., & Bischof, W. F. (Eds.). (2014). *Methods of heuristics*. Routledge.
Hesketh, B. (2000). Time perspective in career-related choices: Applications of time-discounting principles. *Journal of Vocational Behavior, 57*(1), 62–84.
Hsee, C. K., & Zhang, J. (2004). Distinction bias: Misprediction and mischoice due to joint evaluation. *Journal of Personality and Social Psychology, 86*(5), 680–695.
Huselid, M. A. (2018). The science and practice of workforce analytics: Introduction to the HRM special issue. *Human Resource Management, 57*(3), 679–684.
Jolls, C., Sunstein, C. R., & Thaler, R. (1998). A behavioral approach to law and economics. *Stanford Law Review, 50*(5), 1471–1550.
Kaufman, B. E. (2014). The historical development of American HRM broadly viewed. *Human Resource Management Review, 24*(3), 196–218.
Kehoe, R. R., & Wright, P. M. (2013). The impact of high-performance human resource practices on employees' attitudes and behaviors. *Journal of Management, 39*(2), 366–391.
Krosgaard, M. A., Brodt, S. E., & Whitener, E. M. (2002). Trust in the face of conflict: The role of managerial trustworthy behavior and organizational context. *Journal of Applied Psychology, 87*(2), 312.
Lawler, J. J., & Elliot, R. (1993, June). Artificial intelligence in HRM: An experimental study of an expert system. In *Proceedings of the 1993 conference on computer personnel research* (pp. 473–480).
Lazear, E. P., & Shaw, K. L. (2007). Personnel economics: The economist's view of human resources. *The Journal of Economic Perspectives, 21*(4), 91–114.
Lepak, D. P., & Shaw, J. D. (2008). Strategic HRM in North America: Looking to the future. *The International Journal of Human Resource Management, 19*(8), 1486–1499.
Malik, A., Budhwar, P., & Kazmi, B. A. (2022). Artificial intelligence (AI)-assisted HRM: Towards an extended strategic framework. *Human Resource Management Review, 33*(1), 100940.

Malik, A., Srikanth, N. R., & Budhwar, P. (2020). Digitisation, artificial intelligence (AI) and HRM. In J. Crawshaw, P. Budhwar, & A. Davis (Eds.), *Human resource management: Strategic and international perspectives* (3rd ed., pp. 88–110). Sage, chapter 4.

Martinez-Gil, J. (2014). An overview of knowledge management techniques for e-recruitment. *Journal of Information & Knowledge Management, 13*(02), 1450014.

Meijerink, J., Boons, M., Keegan, A., & Marler, J. (2021). Algorithmic human resource management: Synthesizing developments and cross-disciplinary insights on digital HRM. *The International Journal of Human Resource Management, 32*(12), 2545–2562.

Miller, D. T., & Ross, M. (1975). Self-serving biases in the attribution of causality: Fact or fiction? *Psychological Bulletin, 82*(2), 213–225.

Mills, S., Costa, S., & Sunstein, C. R. (2023). *The opportunities and costs of AI in behavioural science*. Available at SSRN 4490597.

Nickerson, R. S. (1998). Confirmation bias: A ubiquitous phenomenon in many guises. *Review of General Psychology, 2*(2), 175–220.

Paserman, M. D. (2008). Job search and hyperbolic discounting: Structural estimation and policy evaluation. *The Economic Journal, 118*(531), 1418–1452.

Pedersen, T., & Johansen, C. (2020). Behavioural artificial intelligence: An agenda for systematic empirical studies of artificial inference. *AI & Society, 35*(3), 519–532.

Ployhart, R. E., & Harold, C. M. (2004). The applicant attribution-reaction theory (AART): An integrative theory of applicant attributional processing. *International Journal of Selection and Assessment, 12*(1–2), 84–98.

Prikshat, V., Patel, P., Varma, A., & Ishizaka, A. (2022). A multi-stakeholder ethical framework for AI-augmented HRM. *International Journal of Manpower, 43*(1), 226–250.

Radonjić, A., Duarte, H., & Pereira, N. (2022). Artificial intelligence and HRM: HR managers' perspective on decisiveness and challenges. *European Management Journal*. ISSN 0263-2373.

Raisch, S., & Krakowski, S. (2021). Artificial intelligence and management: The automation–augmentation paradox. *Academy of Management Review, 46*(1), 192–210.

Rasmusen, E. (2001). *Games and information: An introduction to game theory* (3rd ed.). Blackwell.

Ryan, A. M., & Ployhart, R. E. (2000). Applicants' perceptions of selection procedures and decisions: A critical review and agenda for the future. *Journal of Management, 26*(3), 565–606.

Schoenfelder, T. E., & Hantula, D. A. (2003). A job with a future? Delay discounting, magnitude effects, and domain independence of utility for career decisions. *Journal of Vocational Behavior, 62*(1), 43–55.

Sekiguchi, T., & Huber, V. L. (2011). The use of person–organization fit and person–job fit information in making selection decisions. *Organizational Behavior and Human Decision Processes, 116*(2), 203–216.

Shelley, M. K., & Omer, T. C. (1996). Intertemporal framing issues in management compensation. *Organizational Behavior and Human Decision Processes, 66*(1), 42–58.

Thaler, R. H., & Benartzi, S. (2004). Save More TomorrowTM: Using behavioral economics to increase employee saving. *Journal of Political Economy, 112*(S1), S164–S187.

Tversky, A., & Kahneman, D. (1974). Judgment under uncertainty: Heuristics and biases: Biases in judgments reveal some heuristics of thinking under uncertainty. *Science, 185*(4157), 1124–1131.

van Rijmenam, M., & Logue, D. (2021). Revising the 'science of the organisation': Theorising AI agency and actorhood. *Innovation, 23*(1), 127–144.

Yang, Y., Shamim, S., Herath, D. B., Secchi, D., & Homberg, F. (2023). The evolution of HRM practices: Big data, data analytics, and new forms of work. *Review of Managerial Science*, 1–6.

Zhang, Y., Xu, S., Zhang, L., & Yang, M. (2021). Big data and human resource management research: An integrative review and new directions for future research. *Journal of Business Research, 133*, 34–50.

The manufacturer's authorised representative in the EU is Springer Nature Customer Service Centre GmbH, Europaplatz 3, 69115 Heidelberg, Germany. If you have any concerns regarding our products, please contact ProductSafety@springernature.com

Printed and bound by CPI Group (UK) Ltd, Croydon, CR0 4YY
30/09/2026
02225562-0005